EDUCATION IN A COMPETITIVE AND GLOBALIZING WORLD

NEW DEVELOPMENTS IN SCIENCE EDUCATION RESEARCH

EDUCATION IN A COMPETITIVE AND GLOBALIZING WORLD

Additional books in this series can be found on Nova's website under the Series tab.

Additional e-books in this series can be found on Nova's website under the e-book tab.

EDUCATION IN A COMPETITIVE AND GLOBALIZING WORLD

NEW DEVELOPMENTS IN SCIENCE EDUCATION RESEARCH

NATHAN L. YATES
EDITOR

nova publishers
New York

Library of Congress Cataloging-in-Publication Data

ISBN: 978-1-63463-914-9

Published by Nova Science Publishers, Inc. † New York

CONTENTS

PREFACE

This book presents leading research and new developments in science education. Topics include developing a research programme in science education for gifted learners; research and innovation on learning and about advertising in science education; collaborative curriculum development of a teaching and learning module on bionics based on innovative ICT technology; the role of student questioning in university science classes; scaffolding inquiry-based science and chemistry education in inclusive classrooms; Frederick Accum (1769-1838) and the application of chemistry to social problems; the exhibitions of the natural history paleontology area at the British natural history museum based on argumentation; dynamic visualizations in the classroom; and navigating the scientific landscape via a writing across the curriculum program.

Chapter 1 – An urgent issue in science education is better meeting the needs of those learners who are working at levels of attainment well above those of their peers. Variously known as the gifted, talented, or highly able, these learners are not sufficiently challenged by mainstream science provision, and so may lose enthusiasm and engagement from science. This is not only an issue for the students themselves, but also for the global community that faces environmental, medical and other challenges that will rely upon advances in science. Whilst there has been a considerable amount of work on offering science enrichment provision in many countries, the nature of science education needed for the gifted has not been a main focus on research within the field. The chapter sets out the basis of a research programme to respond to this deficiency. The key concepts and issues are discussed, and 'hard core commitments' of such a programme are proposed. The characterisation of the research programme allows the proposal of an outline 'positive heuristic' for the programme setting out the key directions for future research.

Chapter 2 – In our everyday lives we are surrounded by a huge volume of advertising, which displays a dizzying amount of variety. Modern citizens need an ability to critically analyze this type of media offerings. Advertising is addressed in different school subjects, primarily the humanities and social sciences. In these subjects, the main foci of education which examines advertising include the areas of advertising-specific language, images, illustrations, the use of stereotypes, various strategies of persuasion, etc. But advertising also contains scientific information, which is used either explicitly or implicitly. Scientific claims are made in order to bolster the credibility of ad campaigns and to underline the effectiveness of certain products. Understanding the use (and abuse) of such information is an important skill for critical consumers. However, learning through and about advertising has yet to play any significant role in modern science education up to the present. This chapter describes a project aimed at researching and finding innovative learning approaches to advertising education in the science classroom. It covers both learning through and learning about advertising. The current chapter contains a theoretically justified answer on the topic, including both research on teacher's perceptions of the issue and evidence-based development of the curriculum and corresponding pedagogies. It also addresses the implications for innovating science education with regard to a higher level of critical media literacy.

Chapter 3 – Bionics is a topic in which learning about modern applications of science with reflection on current changes in science and technology can be combined. However, the availability of innovative teaching and learning materials on bionics is still rather limited. This chapter addresses a curriculum development, classroom-based research project which follows the model of Participatory Action Research in science education. In a two-year timespan, two university science educators and four high school teachers collaboratively developed, tested and revised a lesson plan for teaching bionics in lower secondary science education. The central pedagogical innovation was to incorporate all teaching and learning activities in a multimedia-based learning environment. This environment combines different sorts of media and activities such as explanatory texts, learning videos, physical models, inquiry-type experiments, and assessment tasks. The classroom scenario with the corresponding learning environment is described. A reflection upon the experience gained during the collaborative process of curriculum development is also provided. Selected findings are discussed.

Chapter 4 – Students' questions play a crucial role in the learning process since questioning lies at the heart of scientific inquiry and meaningful learning. In the last decades several authors have emphasized the value of students' questions in science learning. A series of studies place the responsibility of questioning onto students rather than their teachers, and indicate that this benefits the quality of science teaching and learning, including at university level. In this chapter we propose to reflect on students' questions in university science classes, considering its relevance for learning and exploring questioning patterns associated to student learning styles and student gender, two topics that remain underexplored.

Chapter 5 – Since the ratification of the UN Convention on Rights of Persons with Disabilities inclusion is on the agenda. In science education inquiry-based learning is, among others, recommended to deal with the diversity of students in a classroom. However, teachers struggle with the implementation of inquiry-based science education and view heterogeneity rather as a burden than an asset. Therefore, this research project is dedicated to the teachers' scaffolding teaching science in inclusive classes. Specifically, within the framework of a qualitative case study, the focus lies upon inquiry-based learning environments with different levels of openness and structuring and how they can be orchestrated to deal with diverse learning needs of all students in one classroom. The research study is conducted in cooperation with an inclusive urban middle school (grade 5 to 8), where students with and without special educational needs learn together.

Initially, the article introduces the inclusive school with its specific way and challenges – not only the science classes – as inclusion is holistic school development. Subsequently, two science formats are particularly focused and contrasted: guided inquiry-based chemistry lessons as well as the format *Lernwerkstatt* (German for workshop center), an open inquiry setting. The aim of the *Lernwerkstatt* is that students work self-dependently on their own scientific questions innervated by carefully arranged phenomena and materials.

Using participant observation and videography data were collected in two 8th grade chemistry classes and nine *Lernwerkstatt* classes. Additionally, audio recordings of informal conversations with the science teachers are used for analysis, which follows the rules of Grounded Theory by Kathy Charmaz. The systematics of Grounded Theory was chosen as it focuses on social phenomena and processes, e.g., teaching processes which need to be analyzed to outline inclusive science education. The differences in the teachers' scaffolding of the open versus the guided setting will be explained.

Chapter 6 – The purpose of this study is to celebrate the life and work of Frederick Accum as an exemplar of the importance of chemistry in the lives of ordinary people. Frederick Accum was born on March 29th, 1769 in Bückeburg, Germany. His father was a soap boiler and merchant and this may explain his early interest in chemistry. He became an apprentice to the firm of Brande, who had businesses in both Germany and England. In 1793, he left Germany to work as assistant chemist in Brande's London Pharmacy. After resigning from Brande's, he set up in business on his own as a chemical analyst. On 10th May, 1798, he married Mary Ann Simpson, with whom he had eight children and of these only two survived to adulthood.

In 1801 he was proposed by Count Rumford as Assistant Chemical Operator at the Royal Institution. A portrait of Accum as a young man shows him to be very finely dressed in the latest style and the fact that he established himself with influential friends demonstrated considerable social skill. In 1803, Accum published his first book, *System of Theoretical and Practical Chemistry* which was based on Lavoisier's system of nomenclature. Over his lifetime he wrote about a dozen books and many chemical papers.

Also in 1803, Accum resigned from his position at the Royal Institution and instead lectured at the Surrey Institution. At the same time he gave private lectures from his own home and teamed up with another German émigré, Friedrich Winsor, who had a plan for lighting London streets and buildings using gas lighting. Accum was also fascinated by the possibilities and had already experimented on this new technology. He gave evidence on gas lighting to Committees of the House of Commons and of the House of Lords as a chemical expert. His business prospered; he assisted a number of Americans such as Benjamin Silliman through his lectures and by selling them equipment to start university laboratories. In 1812 he became a director of the newly formed Gas Light and Coke Company. However by 1814 there was a severe economic downturn in Britain which harmed his business. In 1816, his wife, Mary died.

Accum, as an analyst, became aware of the problem harmful substances being added to food by manufacturers and wrote a book entitled A Treatise on Adulterations of Food and Culinary Poisons. This received great public acclaim but brought him many enemies as he named companies selling dangerous products. Shortly after the publication of this book, he was accused of damaging books at the Royal Institution Library. The final result was that he was so depressed that he hastily returned to Germany, where he worked as an academic for the remainder of his life; Accum died on 28th June 1838.

Chapter 7 – The aims of this research are to examine an argumentation approach to the layouts of exhibitions and the structure of explanations used in one of the British Natural History Museum's Informal educational institutions. For this purpose, we proceed in accordance with the research questions: first, we have established Checklists for evaluating the criteria of macro and micro perspectives of the layouts of exhibitions and the structure of explanations based on argument. The arguments applied to this research, which focuses on the Natural History Museum's Dinosaur Hall, are as follows: (a) Booth et al.'s (2008) argument for problem-solving and Problem-proposing, as well as (b) Oh's (2012, 2014) scientific reasoning argument, for the scientific methodology, were well applied to Natural History Museum Dinosaur Hall.

Chapter 8 – In this chapter, we describe the reflexive enquiry process in which a team of researchers has engaged as the teacher educators of the pre-service teachers who participated in a university/school collaborative program. This study focuses on a case, which can inform ways of developing intervention strategies for pre-service teacher development. The program has been experimental as a complementary opportunity to the teaching practicum, undertaken under the support of the partnership between the department and the participating school, and sponsored by the local Office of Education.

A dozen pre-service science teachers participated in the overall process of planning, implementing and reflecting during ten sessions of the inquiry-based biology class. The primary role of the research team has been to provide, not only guidance on instructional content and methods, but also opportunities to 'reflect on action' by engaging in critical dialogue with the pre-service teachers throughout the whole process. In this study, cultural historical activity theory (CHAT) was adopted as a lens through which to identify the nature of learning as reflexively constituted through the interactions and tensions within and between two activity systems, namely research activity (the researchers being the subject) and program activity (the pre-service teachers being the subject). Over three years, our goal for developing a dialogical mode of intervention strategies has been formed around two questions: i) How can we define and implement an effective co-reflection strategy for enhancing the quality of collaborative teaching? and ii) How can we define and implement inquiry-based science instruction by using laboratory work? In terms of pre-service science teacher professional development, co-reflection strategies have become more fruitful for the PSTs to identify and share issues related to science instruction and to improve accordingly. Concerning the role of teacher educators, more dialogue is necessary in order to facilitate PSTs to think towards the fundamental objectives of science instruction beyond the management of the program. This case study shows how to become more reflexive teacher educators through gaining insight into its own culture and that of the PST community, and ascertaining a third space for communication and collaboration.

Chapter 9 – The chapter presents a study aimed at exploring the advantages of dynamic visualization technology for the development of better understanding of molecular processes. As part of the TELS project (Technology Enhanced Learning in Science), a learning module has been designed in which high school chemistry students conducted virtual experiments with dynamic molecular visualizations of solid, liquid and gas. They interacted with the visualizations and carried out inquiry activities to make and refine connections between observable phenomena and atomic level processes related with phase change. The explanations that students entered in response to pre/post assessments items have been analyzed using a novel scale for measuring the level of molecular reasoning. Results indicated that from pre-test to post-test, students made progress in their level of molecular reasoning, and were better able to connect intermolecular forces and phase change in their explanations. The chapter presents the results through the lens of three improvement patterns: the *extend pattern*, in which more students were able to use an advanced molecular reasoning after interacting with the dynamic molecular models; the *converge pattern*, in which fewer students gave explanations without referring to the molecular point of view after interacting with the dynamic models; and the *shift pattern* combining the previous two, to reach an even higher level of molecular reasoning. The chapter analyses these patterns using the 'ladder of molecular reasoning' metaphor, and discusses how this adds to our understanding of the benefits of interacting with dynamic molecular visualizations.

Chapter 10 – This paper provides insight into issues related to a university-based Writing Across the Curriculum (WAC)/Communication Across the Curriculum (CAC) initiative in the sciences. It focuses specifically on the dialectical implications of cross disciplinary collaboration, the key role played by research findings emanating from this intervention in informing writing praxis in general, and in the sciences in particular, and specific concerns and challenges related to program sustainability. The experiences of other institutions employing a WAC initiative serve as a basis for comparison, support and direction as do the assertions of researchers, experts and practitioners of WAC pedagogy and scientific communication. It is anticipated that the issues highlighted in this paper will serve to inform similar initiatives.

In: New Developments in Science Education Research
Editor: Nathan L. Yates
ISBN: 978-1-63463-914-9

Chapter 1

Developing a Research Programme in Science Education for Gifted Learners

Keith S. Taber*
University of Cambridge Faculty of Education, UK

Abstract

An urgent issue in science education is better meeting the needs of those learners who are working at levels of attainment well above those of their peers. Variously known as the gifted, talented, or highly able, these learners are not sufficiently challenged by mainstream science provision, and so may lose enthusiasm and engagement from science. This is not only an issue for the students themselves, but also for the global community that faces environmental, medical and other challenges that will rely upon advances in science. Whilst there has been a considerable amount of work on offering science enrichment provision in many countries, the nature of science education needed for the gifted has not been a main focus on research within the field. The chapter sets out the basis of a research programme to respond to this deficiency. The key concepts and issues are discussed, and 'hard core commitments' of such a programme are proposed. The characterisation of the research programme allows the proposal of an outline 'positive heuristic' for the programme setting out the key directions for future research.

Introduction

Whilst 'gifted education' has certainly been an area of considerable activity for some decades (Dai, Swanson, & Cheng, 2011; Ziegler & Raul, 2000), there has been much less attention to the particular needs of gifted learners within the science education research community than would seem to have been merited. Yet there are strong reasons to consider this an area that should be made a priority within the field (Taber, 2007b). This chapter sets out the rationale for encouraging research in this area, and offers - if not a manifesto, then at least - a sketch of a research programme (Lakatos, 1970) into science education for the gifted.

* kst24@cam.ac.uk.

It is important to acknowledge that there is no clear consensus on just which learners should be considered as gifted in science, and this is clearly one issue that can impede research. Moreover, some working in education would reject the notion of giftedness as useful either on empirical grounds (it is difficult to unambiguous identify gifted and non-gifted learners) or principled grounds (that the whole notion of some, but not all, people being 'gifted' is unacceptable). Despite this, teachers in many national contexts are expected to provide for gifted learners (as defined and identified in their local contexts), and regardless of acceptance or rejection of *the label* there are good reasons for considering there to be a major issue within science education regarding the differentiation of educational provision for those learners that variously might be recognised as precocious, advanced, highly achieving, of high potential or aptitude in science, etc. (Department for Education and Skills, 2002; Riley, Bevan-Brown, Bicknell, Carroll-Lind, & Kearney, 2004; The National Strategies, 2008).

These points will be picked up later in the chapter, but the general issue is that *any learner will only make good progress in a subject if learning opportunities are matched to their current state of readiness*. This can be seen by considering extreme cases. Consider two learners, both of whom are motivated to learn, and have good study habits: (i) a middle school student being introduced to the general principle that matter is quantised (that apparently continuous matter can be understood to be made up of tiny, somewhat discrete, entities such as molecules and ions), and that the bulk properties of materials rely on their particular structures at a submicroscopic scale; and (ii) a doctoral student undertaking research into the electronic structure of organic conductors at a molecular scale. As a kind of gedankenexperiment we might imagine that these two learners are swapped (perhaps by inquisitive alien educational researchers, or more likely due to some administrative error in an information management system) and consider the learning about molecular structure that would take place.

It seems pretty obvious that the middle school child would not have the background knowledge and understanding to make any headway with the PhD project (Ausubel, 2000). This need not imply the same student does not have potential to one day undertake a PhD in this area (they might), but they are certainly not ready for this yet. Bruner (1960) famously talked of how any learner could be taught any topic in some 'intellectually honest' form - the middle school's curriculum is designed to offer such an intellectually honest simplification of the topic for students at *that* stage in their learning and should offer the 'optimal level of simplification' (Taber, 2000): simplified enough to be meaningful to learners, yet honest enough to the subject matter to support further effective progression in learning. By contrast, the hypothetical PhD project would be set up to reflect an enquiry challenge, at the forefront of knowledge, for someone who has already successfully completed significant advanced learning about the topic.

Similarly the research student is unlikely to develop her understanding of her topic from attending the middle school class, no matter how well informed and skilled the class teacher might be. The doctoral student is certainly well equipped to *understand* the lesson - but the middle school class curriculum does not provide opportunities for further learning about the topic as the content is familiar and the level of treatment well below that level at which the student is working in her thesis project.

These imaginary students are displaced, rather than being natural classmates, and so our gedankenexperiment offers extreme cases, and real classes of students would not normally include such a wide spread of learners. None-the-less, every class - even in a selective education system - is a group of individuals: each with unique background knowledge, different strengths and weaknesses; different interests and aspirations and so forth. In just about any real class *some* of the students will have mastered material others would struggle with; will know things that others were ignorant of; will be skilled in areas others lack skill in; and will hold idiosyncratic alternative conceptions not shared by their peers (Taber, 2014d). If we accept that optimal instruction is based on a match between teaching and the students' current state of learning (Taber, 2011), then differentiation is an issue in every classroom.

On many (if not all) matters that might be educational objectives in our classes - understanding photosynthesis; planning a fair test; using calibrated glassware - there is likely to be a spread of proficiency, and in many cases this is likely to approximate the familiar bell-curve shape of the Gaussian (normal) distribution. In such a distribution, most people score quite near the mean, whilst a few outliers may have scores considerably lower or higher than that mean. For those at the extremes a lesson pitched at the mean will often be problematic. For those most unprepared to work at that level (akin to our imaginary middle school student transported to doctoral studies) there will be little chance of achieving in class without considerable additional support. For those already performing very highly on the relevant indicators (akin to our imaginary research student placed in a middle school classroom), the material met in class is likely to be well-known and the tasks set too easy, and there will be little challenge. This can lead to boredom (Diezmann & Watters, 1997; Gallagher, Harradine, & Coleman, 1997), disengagement with the subject, and - at the very least - a lost opportunity to develop personal understanding and learn anything substantially new.

This is true, and should not be tolerated, in any educational sphere. In science education there is a risk that the most able learners fail to find sufficient challenge in science and look for intellectual satisfaction elsewhere - perhaps in the humanities and social sciences which by their nature more easily offer potential for differentiation because the subject matter (at least as met in schooling) more readily opens itself to students finding potential areas for extending themselves than is usually the case in traditional science courses. This is a generalisation of course, and does not acknowledge the skills of the teachers concerned. It is certainly not *necessarily* the case in science: enquiry work is one area where (when done well) there is considerable scope for learners to show individual flair and imagination (Taber & Cole, 2010; West, 2007); and teaching about the nature of science offers scope for the kinds of open ended intellectual engagement that gifted learners often find in subjects such as history and social studies (Taber, 2007c). Studying socioscientific issues (Sadler, 2011) may be challenging for many learners who excel in science (Levinson, 2007). Indeed, adopting a more 'history-and-philosophy-of-science-informed' approach to teaching about science (e.g. emphasising the nature and role of models and theories rather than letting students see science as factual) could offer both a more authentic reflection of science itself, and make the subject more appealing to many learners (Taber, 2007a).

It is clear that in many national contexts a good many learners become disenchanted with science in the secondary years, and whilst it is healthy that not all students have the same favourite subjects and aspire to the same career, it is commonly suggested that too many of those students with real ability and potential in science are choosing *not* to study science in advanced courses; *not* to major in science; *not* to seek to become scientific researchers; and *not* to look to work in a scientific career. Yet having a strong flow of capable and enthusiastic young people, and indeed a share of those with exceptional potential, is not a luxury in a world beset by problems that science and technology can address. There are certainly some important issues where science and technology can make a major difference to our lives:

- Major outbreaks of fatal infections that medical science cannot yet readily cure;
- Regular food shortages in some parts of the world;
- Loss of natural habitat and biodiversity;
- Pollution from modern industry and consumer habits;
- Reliance on non-renewable, and highly polluting, power sources;

And so forth…

Science has potential to help tackle disease, provide 'greener' manufacturing, offer new fabrics which are better wearing and more comfortable and even able to adapt to different environmental conditions, provide new digital technologies, clean up pollution, develop new crops with disease and drought resistance, etc. Science will continue to do these things - but is likely to make better progress if a fair proportion of students with the 'brightest minds' opt to work in scientific areas. If *too many* of those learners that we might judge gifted in science find other areas - such as perhaps business and finance, or media, or advertising - to be more attractive and satisfying than science then progress on major issues effecting millions of lives (and perhaps ultimately human survival) will be slower.

So whilst there are certainly some challenging and even sensitive issues regarding how to best understand giftedness in science, there is also a very real concern that science educators in many national contexts need to do more to make sure that science learning is intellectually challenging and satisfying for the most able learners. The purpose of this chapter is to offer an outline of a research programme that could act as an organising schema for this area of research across educational phases and national contexts. First however, the notion of research programmes is explained.

Science Education As a Field or Fields

Science education is now widely recognised as a substantive research field (Fensham, 2004), or possibly even set of fields (Gilbert, 1995), and is mature enough to have developed a suite of prestige journals and a range of high profile associations. Moreover, science education is an active international field with traditions of work across different continents (Taber, 2012a). The increasing maturity of science education as a field is reflected in two major handbooks now being in their second editions (Fraser, Tobin, & MacRobbie, 2012; Lederman & Abell, 2014).

The nature of science education itself as an area of scholarship is less straightforward. In particular the disciplinary background of people working in the field is diverse: with some workers trained as natural scientists who have moved into pedagogic research; some who have moved from professional careers as school teachers before becoming research active (and many but not all of these will be science graduates); and others who have backgrounds in those disciplines on which education is founded (e.g. psychology, sociology), but who have developed lines of research within the context of science education whilst retaining a professional identify as a psychologist or as a sociologist etc.

By its nature science education is not part of natural science, although many of those active in science education research - and particularly in discipline-specific subfields, such as chemistry education research (Taber, 2013b) - are based in university science departments. Education is more of an applied social science, and when selecting methodologies appropriate for answering educational questions researchers usually need to adopt approaches developed in the social sciences rather than the natural sciences (Taber, 2014b).

However there are different understandings of what is meant by 'science', and whilst education is not a natural science, and indeed sometimes encompasses arts and humanities based research methods as well as social science approaches, it is still possible that research in education can be considered to be *scientific* (National Research Council Committee on Scientific Principles for Educational Research, 2002).

RESEARCH PROGRAMMES IN SCIENCE EDUCATION

The question of the demarcation of science from other forms of activity has been discussed by a number of philosophers of science (e.g. Lakatos, 1974/1981; Popper, 1934/1959, 1974/1992). One particularly useful description of scientific activity was developed by Imre Lakatos (1970), who described the nature of *scientific research programmes*. For Lakatos a research programme had particular features that gave it its 'programmatic' (rather than serendipitous) nature, and which allowed it to be evaluated to determine whether or not it was 'progressive'. Adherence to a research programme that was not progressive, rather than seeking or developing a more promising alternative, was not scientific - and this was used as a criterion for claims that such programmes as Marx's notion of history or Freud's psychoanalytical programme fell short of being scientific (i.e. progressive).

Lakatos's model of science addressed some of the problems of a naive conception of science as able to produce absolute knowledge that is verified by empirical enquiry. Whilst the very essence of science is to be guided by empirical evidence, it is widely accepted that this is not unproblematic. For one thing, theories are generally underdetermined by data (Riggs, 1992). That is, no matter how well the empirical evidence fits one theory, it is quite possible to conceive of alternative theories that would offer just as good a fit. As a visual analogy - consider a line graph based on a finite number of data points. There may be an obvious line of best fit, but there are always other more complex functions that would also fit the data. The scientific equivalent of picking the best-fit curve relies on adopting values that are to some extent at least metaphysical. For example, William of Ockham's principle (known sometimes as Ockham's or Occam's razor) that the most straightforward explanation with the least auxiliary features is to be preferred is commonly used. This seems a sensible heuristic - but it cannot assure the right choice. It seems more likely that our imaginary swapped students find themselves in the wrong place due to an administrative error than because of alien intervention - but occasionally what seems unlikely may turn out to actually be the case.

Another problem in science is that very few scientific experiments can be considered to be a test of simply the hypothesis being investigated. Most such experiments are based on propositions that are logically of the form 'if…and if…and if…and if…and if…then…'. The 'and ifs' can refer to assumptions from existing theory that is widely accepted ('hard core' commitments within a particular scientific tradition) or understanding of the instrumentation (often very complex in modern science). As we have seen above, a positive result underdetermines any particular conclusions, and largely under the influence of Popper (1934/1959) science is often now often described as hypothetical-deductive, working on a principle of producing conjectures and refutations (Popper, 1989). In Popper's model science proceeds by excluding hypotheses that are demonstrated false: i.e., falsificationism. Yet an experiment cannot lead to the absolute rejection of a discrete conjecture - only to a rejection of the conjunction of 'if…and if…and if…and if…and if'. These 'and if's may include things such as 'if the sample was of high purity and not contaminated' and 'if the power supply remained within close parameters' and 'if the technician added just the right amount of reagent' and 'if the observer accurately counted the damaged cells in the field of view' and 'if we selected an area of the habitat that was typical of an animal's range' and so on.

Lakatos recognised that despite the popular accounts sometimes met in school books, no single experiment can ever really be considered critical, and that scientists will often happily quarantine anomalies (results they were not expecting and which cannot - yet - be explained from within the theoretical framework adopted), that according to a simple epistemology should 'refute' their theory or model, and carry on regardless. Despite this appearing to be illogical from a simplistic falsificationist perspective, Lakatos recognised it made sense in a situation where any individual result should never be taken as an absolute basis for supporting or undermining a particular theory. Rather it is rational to put aside ('quarantine') anomalies for the moment, as long as the overall research programme can still be considered to be progressive, i.e. is still making progress. Clearly evaluating research programmes requires more expansive judgements (Lakatos, 1970): i.e., is theory still being developed (beyond *ad hoc* patching-up of failures to match results) and still suggesting fruitful avenues of empirical investigation?

Arguably, science education is a scientific field *to the extent* that those working in the field are adopting progressive research programmes. One area of work that has been identified as a research programme in science education (Gilbert & Swift, 1985) was based upon Piaget's (1970/1972) genetic epistemology and looked to inform curriculum and teaching using Piaget's notion of children's development occurring in stages that afforded different cognitive abilities (Bliss, 1993). This programme has supported extensive work on developing teaching aimed at 'accelerating' cognitive development by offering learners the contexts for moving on to the stage of (what Piaget referred to as) 'formal' operations needed to appreciate the abstract nature of scientific concepts and theories (Adey, 1999).

An alternative perspective that has been widely recognised as a major tradition in science education, a paradigm or research programme (Driver & Easley, 1978; Gilbert & Swift, 1985; Solomon, 1993; Taber, 2006, 2009b), has been based in constructivist ideas on how learners come to knowledge and how this should inform teaching. This constructivist programme drew upon general features of Piaget's work, but considered how learning occurs in particular concept areas drawing upon the resources available to a learner - such as prior learning (Taber, 2014c) - rather than accepting that the learner operates with domain-independent general levels of cognition. This area of work included a vast research effort to explore the nature of learners' thinking prior to being taught science topics (Duit, 2009), and has led to the reporting of many 'misconceptions', 'alternative conceptions', 'intuitive theories', 'alternative conceptual frameworks' and the like.

Indeed this area of work become so expansive that it faced some criticisms for unduly dominating the field (Solomon, 1994), to the extent that at one point it seemed to be by far the most visible area of activity in science education research. This is no longer the case, and for example in recent years there has been the development of an alternative research programme which is based upon cultural-historical and socio-cultural theory (Smardon, 2009), and which is increasingly important in science education (Fraser et al., 2012; Taber, 2014a). The physicist and historian/philosopher of science Thomas Kuhn (1996) understood alternative paradigms operating in a field to be in competition, and from that perspective the shift in interest among researchers from the constructivist programme to the sociocultural programme (like that from the Piagetian to the constructivist programmes) could be seen as a kind of paradigm shift. However, in Lakatos' model several different alternative research programmes may continue to be progressive in a field. Given the complexity of factors influencing learning, it seems sensible that a field such as science education can support complementary research programmes with different foci - although one might consider the ideal that these would at some point in the future find ways to coalesce into a single programme.

Indeed one conceptualisation of the constructivist research programme as concerned with the contingent nature of learning in science would admit general features of the Piagetian programme (without treating domain-independent stages as an absolute) and aspects of sociocultural perspectives - seeing any distinction between 'personal constructivism' and 'social constructivism' as a choice of focus for particular studies rather than incommensurable conceptual frameworks (Taber, 2009b).

THE FEATURES OF RESEARCH PROGRAMMES

Lakatos (1970) considered that research programmes could be characterised as having key features, including what he referred to as a hard core, a protective belt, and heuristics to guide research. The hard core was central to the identity of the programme. From his study of historical cases, Lakatos considered that a scientific research programme forms around a set of specific assumptions that are established at the outset, and which act as the (intellectual) starting points for the programme.

A research programme's hard core comprises a set of intellectual commitments that are so central to the programme that they are unquestioned within the programme. A naive notion of science would suggest that everything should remain open to questioning: in Lakatos's model the rejection of a hard core assumption is certainly possible, but in effect means the abandonment of that particular research programme. As an example from science, we might consider a Lamarkian research programme to assume that acquired characteristics can be inherited, and a decision to reject that premise (e.g. to adopt a Darwinian model) means a rejection of the Lamarkian programme.

Piaget's research programme (1970/1972) has as part of its hard core that all children progress through the same series of basic stages of cognitive development, in the same specific order, and that this explains important domain-independent aspects of a child's intellectual performance. Someone working in that programme who came to the view that this basic premise was inconsistent with the empirical evidence would need need to abandon the programme. In such cases it is perfectly possible for a researcher to join (or develop) a new programme that inherits some of the commitments of the abandoned programme (Taber, 2009b), but substantive change in the hard core assumptions implies the development of a new programme.

Lakatos suggested however that those working within, and committed to, a research programme did not readily question hard core assumptions: as normally these were insulated by the 'protective belt' of theory that is developed around the core, and which could be considered to offer 'refutable variants' consistent with the commitments in that hard core. Moreover, unlike in a naive falsificationalist notion of science (where data that appear to be evidence against a hypotheses should lead to its rejection), Lakatos (1970) suggested that in real scientific research programmes scientists would be prepared to quarantine inconvenient results - not to deny them, but to recognise them as potential problems for the programme that should be considered as puzzles that *in time* need to be addressed. A programme that accumulated increasing numbers of such quarantined anomalies that were *not* being tackled would be an indication of a programme that was no longer 'progressive' (but rather was 'degenerative'), and so less worthy of scientists' investment of resources.

A Progressive Research Programme from Science Education

As an example, an analysis of seminal studies from the constructivist tradition in science education (Driver & Easley, 1978; Driver & Erickson, 1983; Gilbert, Osborne, & Fensham, 1982; Gilbert & Watts, 1983; Osborne & Wittrock, 1983), suggested a research programme existed based around a series of key commitments that had been adopted by researchers. These were (but not necessity in the phrasing of the original authors):

- Learning science is an active process of constructing personal knowledge
- Learners come to science learning with existing ideas about many natural phenomena
- The learner's existing ideas have consequences for the learning of science
- It is possible to teach science more effectively if account is taken of the learner's existing ideas
- Knowledge is represented in the brain as a conceptual structure
- Learners' conceptual structures exhibit both commonalities and idiosyncratic features
- It is possible to meaningfully model learners' conceptual structures (Taber, 2009b, p. 124).

The constructivist programme has been so influential (and arguably successful) that few people working in science education today would doubt most of these premises - even if they might quibble over the precise wording. One potential criticism of this characterisation here is that it seems to have a strong focus on the individual learner ('personal constructivism') - however, social aspects of teaching and learning are not excluded and indeed the reference to teaching science acknowledges that learning usually takes place in a social (e.g. classroom) context. However, some scholars go beyond stressing the role of social and cultural factors in learning to suggest that knowledge should be seen as distributed across a community (Collins,

2010) rather than located within individual minds - and someone with that commitment could not be considered to be working within a research programme having hard core personal constructivist premises (such as 'knowledge is represented in the brain as a conceptual structure').

The heuristics of a research programme set out directions for developing the programme. Any one of the premises above suggests empirical questions (what are the mechanisms of personal knowledge construction? what factors influence these processes? etc.) and those that seem to offer promising lines of work form a positive heuristic. Indeed, an analysis of this area of work suggests that each of the hard core assumptions above suggests lines of enquiry which researchers have indeed adopted - making (to date) various degrees of progress (Taber, 2009b). During this work a wide range of theoretical ideas have been developed around the hard core (making up the programme's protective belt).

What Lakatos (1970) called the negative heuristic refers to the taken-for-grantedness (within a programme) of not setting out to test the premises. So, for example, someone working within a research programme developed from Einstein's theory of special relativity would not undertake research to measure how the speed of light varies with an observer's state of motion - as the invariance of the speed of light is a hard core commitment of the programme. A researcher working in the constructivist programme discussed above would not set out to develop a general model representing how all 14 year olds studying in the English education system conceptualise energy - as such a general model is inconsistent with the assumption that each learner has a somewhat unique conceptual structure (i.e. 'learners' conceptual structures exhibit both commonalities and idiosyncratic features'). A researcher *might well* develop a model which represented *the commonalities* of a class of learners in relation to a particular concept area (e.g., Taber, 1998), as the programme includes a commitment to such commonalities: but the notion that a detailed model of a conceptual structure would accurately represent the conceptual structures of each member of a general class of learners is excluded by the hard core of the programme.

The other key aspect of Lakatos's model of research programmes is what he termed the 'protective belt' of auxiliary theory that develops around a programme's hard core. Under the positive heuristic the programme develops a body of theory that is consistent with the hard core commitments. However, as suggested above, any theory is underdetermined by available evidence, and different alternative auxiliary theories may be consistent with any particular hard core. For this reason, such theories are described as refutable variants, as they are open to being rejected and replaced in the light of new evidence. The term protective belt suggests that these theories act to protect the hard core because whenever empirical evidence seems to be inconsistent with the current state of the programme's theoretical apparatus it is possible to adjust the apparatus by modifying the protective belt and leaving the hard core untouched. (That is the '...if...' of a conjunction of 'if...and if...and if...and if...and if' is protected because one of the '...and if...' clauses is sacrificed instead.)

This suggests of course that any set of hard core assumptions can be retained by sufficient adaption of auxiliary theory: and this reflects the conservative nature of much human thinking (Nickerson, 1998). Lakatos was aware of this, and he was not advocating a science that lacked the scepticism usually considered essential to scientific progress. This is why he suggested criteria for distinguishing a progressive (and so scientific) programme where over time the body of theory comes to offer a more coherent explanation for an increasing evidence base from a degenerating programme where preferred commitments are increasingly supported by *ad hoc* and arbitrary patching-up of theory to 'save phenomena'.

The constructivist research programme has seen a wide range of theoretical developments which have motivated and been developed through empirical investigations (Taber, 2009b). As one example: early work on students' ideas in science suggested that 'alternative conceptions' were very tenacious and so difficult to address in teaching. This idea was criticised by those who considered learners' ideas to be in flux, highly contextualised and general labile (e.g. Claxton, 1993). Further research suggested that both views were too simplistic, and that learners' ideas vary considerably in terms of level of commitment, extent of (i) range of application, and (ii) integration within a wider network of ideas (Taber, 2009b). We also now understand much more about the different origins of learners' informal ideas and why they should vary in these ways (Taber, 2014d).

A Research Programme in Science Education for Gifted Learners

The remainder of this chapter will consider the current state of knowledge and understanding about gifted education in science and suggest how this area of work in science education might be developed through an agenda for research conceptualised through a scientific research programme.

Possible Hard Core Assumptions of a Research Programme Exploring Giftedness in Science

It is possible to suggest candidates for the hard core commitments of a research program in science education for gifted learners. For example, the following may be suggested as ontological commitments that underpin this area of work:

1) A student's potential to benefit from teaching depends upon their current characteristics as a learner.
2) Learning depends upon learning opportunities being matched to a student's characteristics.
3) Learners show a wide range of current capabilities within any particular curriculum area, and vary considerably in terms of their potential for development with suitable support.
4) At any age or grade level there will be some students who are considerably further advanced in their science learning than their peers. [These students may sometimes be referred to as gifted in science.]
5) Science (and science subjects) present a broad range of learning challenges such that many students will have an uneven profile of current knowledge, understanding and skills.

6) Science provision needs to offer suitable learning opportunities to allow students to make progress in their learning. (Therefore: science provision needs to offer suitable learning opportunities to allow 'gifted' students to make progress in their learning.)
7) Particular science provision is indicated for those learners in any cohort who are too advanced to find substantive learning opportunities in the standard provision.

These commitments are consistent with the approach taken above in the introductory sections of this chapter. This characterisation of the background to a research programme is not exclusively about giftedness - what becomes clear when the theme is analysed in this way is that most of the underlying notions necessary to develop a scientific research programme into giftedness in science are shared with more general thinking about science teaching and indeed teaching in general.

The purpose of setting out the hard core in this way is both to make explicit what are considered to be essential shared commitments (i.e. does anyone looking to research giftedness in science education actually disagree with any of these points?), but perhaps more importantly to act as a starting point for identifying the questions that are suggested by these assumptions. Such questions, where robust answers are not already available, indicate directions - lines of work - for the research programme (the 'positive heuristic').

IS THERE A NEGATIVE HEURISTIC FOR THIS RESEARCH PROGRAMME?

According to Lakatos' (1970, p. 135) model ('method') of scientific research programmes, such programmes have a negative heuristic that "specifies the 'hard core' of the programme which is 'irrefutable' by the methodological decision of the protagonists". In other words, someone who is working within the programme accepts (for example, in the mooted programme) that "at any age or grade level there will be some students who are considerably further advanced in their science learning than their peers" and would not develop a line of work that was based upon an assumption that all students of a particular age are currently capable of understanding or mastering the same set of concepts.

As the hard core represents strong ontological commitments there is a sense in which it tends to become taken-for-granted by those working in the field. So in practice the negative heuristic need not be a clearly specified set of rules, but can simply operate implicitly in guiding how researchers work. Yet in this particular area of activity, there is arguably a good reason to be explicit about the hard core of the programme. The list of candidate commitments for a hard core above does not seem especially radical, and it seems likely that few working in science education - or indeed with any experience of teaching in any academic area - would strongly disagree with any of these points. Yet a great deal of educational practice seems to be predicated on a very different set of (if often implicit) assumptions - that it is appropriate to set standard curriculum for particular grade levels, and to ask teachers to teach heterogenous, and sometimes widely heterogenous, groups of students following a common syllabus or scheme-or-work. In part this represents traditional practices that have themselves become taken for granted, and in part this recognises the economic realities of public education (i.e. the need for teaching in classes of a size that reflects the teaching staff resources available).

Certainly at school level it is most common to organise classes by age-related grade - that is where the criterion for being in a grade usually relates to birth date rather than prior attainment. This is certainly administratively neater when legal requirements set out the ages of compulsory schooling. Attempts to move away from this system to see grade level in terms of prior academic achievement may be complicated when students have an uneven profile of achievements across curriculum areas, or where precocious intellectual ability is not matched by physical and emotional development: making the learner placed in a class of mainly older children feel out of place socially - or even isolated. Where a largely age related system is modulated by limited advancing of students to higher grades or holding back those not considered to be making progress there is the potential for considerable risk to student self-esteem and self-image when they are kept back with younger pupils (Martin, 2011).

Teaching uses three sets of tools in response to this situation. One is ability grouping - setting, banding etc. within grade levels (where there are enough pupils of the same age group within an institutional cohort to make this possible). The second is differentiation (Stepanek, 1999) - finding ways to differentiate within a class by modifying tasks, or setting tasks at which different levels of achievement are possible. The third tool is enrichment through activities that supplement standard curriculum fare. Each of these approaches has advantages and limitations - and practitioners would benefit from a stronger research base exploring the most appropriate forms of provision for gifted learners in science classes in various contexts and situations (e.g. institutional and curriculum contexts, disciplinary contexts, age/grade levels, etc.)

A Positive Heuristic for the Research Programme

The characterisation of the hard core of a research programme should indicate directions for potentially fruitful research. It is possible to look at the suggested hard core commitments above and consider how they suggest questions such a research programme should seek to answer.

A Student's Potential to Benefit from Teaching Depends upon Their Current Characteristics as a Learner

This assumption is widely accepted in education. When considered as a commitment within a research programme this principle raises issues of *how to best characterise learners to identify their potential to benefit from further teaching*. There is a great deal of material available to teachers on how to identify gifted learners - but much of this appears to link to 'craft knowledge' rather than be based upon sound research. A very common indicator used to define gifted learners is intelligence as measured through IQ tests, but there is a wide literature suggesting this is a much too limited approach (Charlton, 2009; Gilbert & Newberry, 2007; Gould, 1992; Sternberg, 2010). Teachers may instead use classroom experience to simply identify which students complete work quickly and at a high level with little apparent effort. There are also common checklists that offer suggestions of gifted behaviours, such that being considered to demonstrate a selection of the listed behaviours is taken to indicate giftedness (Riley et al., 2004). Such lists usually contain what seen sensible potential indicators, but the origin, definition, and criteria for meeting, the indicators are often something of a mystery.

This all suggests that whilst the general principle (i.e., that a student's potential to benefit from teaching depends upon their current characteristics as a learner) is widely accepted, there is limited robust (i.e. systematically tested) knowledge about how to best characterise learners to predict their potential for further learning - and so there is much room for developing more evidence-based practice in this area.

One very useful theoretical idea that has gained widespread adoption in educational discourse is that of the zone of next (or proximal) development (ZPD), proposed by Vygotsky (1978). Vygotsky distinguished between what a learner could currently achieve unaided (their zone of actual development, ZAD) and what they might achieve with suitable 'scaffolding' from a teacher or more advanced peer (in their ZPD). Vygtosky's argument was not just that development takes place when working beyond the 'comfort zone' of what we can can already do well, but also that within the same class students who appear to have similar ZAD (i.e., seem to be achieving at much the same level in terms of tasks that have been mastered) may have quite different ZPD and therefore need different levels of support (and challenge) to move them on.

Whilst it is possible to find much writing about the idea of the ZPD (and, indeed, different interpretations of this idea) there has been little research in science education to actually demonstrate (or test) Vygotsky's claim and see how it can be applied in practical terms. This is important, as if Vygotsky is correct, students' scores on tests (undertaken usually unaided and individually - that is working in the ZAD) do not necessarily give a helpful indication of what the learners are ready to achieve next with suitable teaching or peer support (working in their ZPD).

The notion of the ZPD might be considered an example of the kind of theoretical content that populates the 'protective belt' of a research programme (Lakatos, 1970). That is, the principle that 'a *student's potential to benefit from teaching depends upon their current characteristics as a learner'* is considered fundamental to, and so irrefutable within, the research programme - but the notion of the ZPD as an effective theoretical tool to characterise learners is what is sometimes termed a 'refutable variant' of the programme. It is considered a *refutable* variant as the notion of the ZPD could be sacrificed without undermining the programme as long as alternative ways of conceptualising how to characterise learners to indicate learning potential could in principle be developed, tested, and adopted.

Learning Depends upon Learning Opportunities being Matched to a Student's Characteristics

Whether the specific idea of the ZPD or some alternative transpires to be the most productive way of characterising students' readiness to benefit from challenging learning opportunities, the second candidate as a tenet for our research programme posits that supporting learning requires a process of matching learning opportunities (and so teaching) to the characteristic of the learners. A clear question for research is *how* to effectively carry out such matching: how can learning opportunities be matched to students' characteristics so as to best facilitate learning. This question potentially has two strands. In part this is an 'in principle' question looking at how methods of matching can be found which 'work'; and in part this is a practical question of viable professional practice. First research would seek to show this can be done - even if this requires considerable resource (e.g. small class sizes, additional instructional materials). Then research is needed to show how this can be incorporated in professional work as part of normal procedures and practice.

That is, in research we often commit a level of resource to the foci of our investigations (particular classes, teachers, students, institutions, lessons, etc.) that is well beyond the level of resource it would be possible to commit in practice (i.e. to all classes, teachers, students, institutions, lessons, etc.). This is often necessary to systematically study a situation, or to iteratively develop instrumentation, innovations, teaching resources, etc. that allow us to achieve objectives.

Ultimately, however, for research to be useful it needs to be applicable. This candidate commitment for our research programme then indicates something more than seeking 'in principle' demonstration that effective matching is possible. Rather, the ultimate aim is to find ways of supporting teachers (and curriculum developers, etc.) in undertaking effective matching of learning opportunities to student characteristics within the resource constraints of normal teaching: large classes, busy teaching timetables, and limited budgets that make the purchase of special diagnostic tools unlikely.

Learners Show a Wide Range of Current Capabilities within any Particular Curriculum Area, and Vary Considerably in Terms of Their Potential for Development with Suitable Support

That learners show a wide range of current capabilities within any curriculum area is familiar to those working in the education system. Indeed this is an inherent assumption behind public examination systems that seek to certify learners across different levels of achievement, especially when certification is used for selection purposes in terms of progression (such as entry to further or higher education; entry into employment).

That learners demonstrate a wide range of current capabilities can be understood in terms of complementary considerations. What students demonstrate need not be their full potential: some learners take demonstrating their capabilities in class or in examinations more seriously than others. So there are issues of motivation for example (Phillips & Lindsay, 2006) - some students are motivated to do well in tests and examinations, to answer the teacher's questions, to do their homework to the best of their abilities and so forth. Other learners may be capable of demonstrating the same levels of achievement, but are not motivated to do so sufficiently to put in a high level of effort - and so achieve less than more motivated peers.

Moreover, current capabilities are dependent (as suggested above) on learning opportunities. A student can only demonstrate those capabilities they have been facilitated to develop - regardless of their original 'potential'. Of course, this will depend to some extent on what kind of capabilities we are referring to. Some students may have the ability to interpret graphs well despite limited learning opportunities - but they are unlikely to be able to, for example, apply the Brønsted-Lowry theory of acids if they have never been given the opportunity to learn about that theory.

There are cultural differences in this area of thinking within education. In what might be thought of as a Western (e.g. Anglo-American) tradition it is common to think of people as having widely varying inherent potential that is met or not according to opportunities and effort. In this tradition being clever or bright (or alternatively 'thick' or 'slow') is a characteristic of the individual that is relatively fixed and dominates attainment. This idea may be anathema in some Asian contexts where it may be assumed that commitment, engagement and application to a task always bring rewards. Of course such patterns should not be generalised to make assumptions about all students from a particular culture (Ryan & Louie, 2007), but none-the-less students transposed to a different educational context may well experience some dissonance in terms of what is widely taken for granted about the causes of academic success. There may also be gender differences in students' own explanations for their academic successes (or failures) in some educational contexts with boys in general more likely to put success down to ability and failure to lack of engagement, and girls more likely to ascribe failure to lack of ability and success to sheer effort. The phrasing of the candidate for a research programme's hard core commitment here (…range of current capabilities…potential for development…) is intended to leave open the extent to which a students' current achievements or their potential for future attainment may be *partially* determined by factors that need to be considered 'fixed' from an educational perspective. This is an important area that deserves more attention within science education - bearing in mind that however extensive or limited the malleable components may be, it is here that teachers can make a big different to their lives of their students.

There is clearly a good deal of evidence already available about the range of capabilities students at different levels demonstrate in relation to various learning objectives. What is less clear is how this can be linked to an understanding of the potential that different learners have to make progression in learning from those starting points given suitable support. There is research suggesting how some learners may be constrained by their level of intellectual development from accessing certain types of science learning objectives (Shayer & Adey, 1981) although such notions are by no means universally accepted (Sutherland, 1992). Teachers will also apply their professional knowledge to make judgements about how to manage progression in learning for different students. Yet there is limited clear detailed guidance on *how to* judge a student's potential for learning with suitable support - for example in different areas of the science curriculum.

This is generally true for all learners, but is perhaps most pressing in the case of the students who might be labelled gifted. Where teachers over-estimate the progress students can make, the results (poor learning outcomes) are clear, and adjustments to teaching can be made. In the case of gifted learners, however, it is often likely that teachers will set work which is considered challenging for the class, and which the gifted learners will respond to successfully. This therefore seems a very positive outcome, as these students have risen to the challenge. This may reflect a well-made judgement about what those students are ready to tackle next. However, it could instead simply indicate that there was limited challenge for the gifted learners in the task, and that these students managed to achieve without making the extent of learning gains that they were capable of. If students have been socialised into expecting to work without special effort or challenge, they may indeed be perfectly content in being given, and responding to, tasks to undertake that they can competently and comfortably complete without having to metaphorically 'break into a sweat' - just being kept occupied with 'busy work' (Moursund, 2006). If students have learnt to be sufficiently rewarded by the external motivators of grades and teacher praise they may lack any insight into how far from their full potential they are being asked to perform.

In the case of individual classes, there is scope for trial and error here: if the gifted learners are readily successful, then teachers can perhaps look to make even greater demands in the next task until a level is found at which they are challenged enough to have opportunities to make real progress. In principle this sounds sensible, and indeed - given the complexity of teaching and learning - good practice is always likely to rely to some degree on this kind of iterative approach, even when supported by a strong research base that provides clear guidance.

However, the reality is that the most gifted learners in many classes are already under-performing in the sense that they have been successfully meeting task demands that were seldom stretching them, given that the gap in *appropriate* teacher expectations between some gifted learners and their classmates is likely to be considerable. This offers a real challenge to the teacher. Some tasks lead themselves to differentiation by outcome. For example, setting an assignment to research the contemporary response to the publication of Darwin's (1859/1968) *On the Origin of Species* is open-ended enough to cover a broad range of abilities (one suspects this has been the topic of more than a few doctoral dissertations). However, there are many topics where there is a significant tension between 'class teaching' and differentiation - especially in contexts where teachers are encouraged to undertake more whole class teaching. So one strategy is to treat a class as a collection of subgroups each of which the teacher works with separately - yet there is much discourse in education suggesting that good teaching is often teacher-led 'direct instruction' (Klahr, 2009; Taber, 2010), and the expectations (of head-teachers/principals, of school inspectors, of parents) of what this looks like may translate to considerable proportions of time when the teacher is working with, and leading discussion of, the whole class (Key Stage 3 National Strategy, 2004). Such expectations may work against effective differentiation techniques. A good teacher can keep a class engaged in an extended discussion - but this may not mean that everyone in the class is both (a) sufficiently understanding the ideas being discussed, and (b) also being challenged enough in their thinking to promote further learning.

Given these considerations, there is a strong need for a research base to find out a lot more about the range of potentials for further learning in different kinds of classes. Without this it is quite possible that norms and expectations about how classes are organised in many education contexts might be necessarily limiting the learning opportunities of some students through teaching that fails to sufficiently challenge their current thinking and scaffold new learning.

At any Age or Grade Level there Will Be Some Students Who Are Considerably Further Advanced in Their Science Learning than Their Peers

The diversity of learners within any cohort or class is clear to all teachers. Effectively pedagogy involves engaging with learners' thinking, and indeed good science teaching is very much an iterative process where the teacher is operating with an (often implicit) model of what the learner already knows and understands, and so pitches information, questions, tasks accordingly - and then uses information about how the student responds to inform modifications of that internal model (Taber, 2013a). This is at the basis of learning that is sometimes referred to as constructivist (Taber, 2011), but which is also sometimes labelled as learner-centred (or even 'progressive'). Unfortunately such terminology is not used in uniform ways in educational discourse - and terms such as constructivist and progressive are seen by some as bywords for ideologically driven approaches that undermine ordered teaching (Kirschner, Sweller, & Clark, 2006). That is misinformed, as constructivist teaching done well is not *laissez-faire*, but is rather teaching that is matched to students' current learning needs (with an optimal level of guidance), and which moves between teacher-led and student-led activities in a planned and structured way (Taber, 2013a).

Anyone teaching in an interactive way (and arguably non-interactive teaching is an oxymoron - and is no more than presenting) soon becomes aware of the diversity of capabilities within a teaching group - even when working with what is labelled - for example - a top set. Indeed if we assume that within any population different capabilities are likely to follow a distribution something like a normal distribution (as suggested above) then we might expect those groups which contain the learners considered most able (and to some extent also those groups which contain the learners considered least able) to show the widest variation. So for example, on an IQ test there will be many scores clustered near the mean, such that a proportion (e.g. 10%) of individuals with scores closest to the mean will have a modest spread of scores. However, the same proportion (e.g. 10%) of individuals at either tail will show a much greater range of scores.

These Students May Sometimes Be Referred to as Gifted in Science

Terms such as gifted, or gifted and talented, are commonly used to refer to these students - although other terms such as highly able may also be used (Cropley & Dehn, 1996; Department for Education and Skills, 2002; Sternberg, 1993). However, there is little agreement on nomenclature or how to define these groups in the literature. In this chapter the label 'gifted' has been adopted as it is commonly used and is widely understood at least in general terms. Part of the work of a research programme will be to clarify and adopt terminology and nomenclature that can be readily used across an international community of researchers.

Science (And Science Subjects) Present a Broad Range of Learning Challenges Such That Many Students Will Have an Uneven Profile of Current Knowledge, Understanding and Skills

Another issue of importance in science teaching is that science is itself quite diverse (Jenkins, 2007). There is certainly commonality across the sciences - at least in terms of a mindset that informs a way of going about science. However, the subject matter of science is extremely diverse (being, in principle, all of nature) and the methodologies adopted are similarly so - with some sciences being highly experimental, some largely based on observation; some being mostly laboratory based, and others relying on fieldwork. Even within a science, there is a considerable range in what is required of a learner.

Chemistry can be considered as an example. Chemistry is a subject that involves a good deal of laboratory bench work, requiring good manipulative skills (as well as an understanding of why and when to use those skills) as well as a diverse body of theory to be mastered. The content of chemistry has traditionally been structured through the main branches of inorganic, organic and physical chemistry (even if such distinctions are slowly being broken down), and each of these has its own 'feel' and nature. Physical chemistry relies a great deal on mathematical modelling, and tends to deal with general principles and theories that can be applied to a wide range of chemical systems. Inorganic chemistry has a very strong link to the periodic table as the basis of a system of classifying substances through what can be a very ordered series of typologies. The mind-set for excelling in physical chemistry is surely somewhat different from that required to master inorganic chemistry. Organic chemistry is in some ways more akin to inorganic than physical chemistry - being organised through classification systems based on actual substances and reactions rather than general theoretical models. But there are still very real differences between these two branches - in part due to the complexity of the range of organic substances that exist (or can be brought into existence). It is quite conceivable that a learner considered gifted in physical chemistry might not show such exceptional ability in organic chemistry - or *vice versa*. Similarly, those who excel in theoretical chemistry may not be especially skilled at the bench. Moving into biochemistry or pharmaceutical chemistry crosses further disciplinary borders.

It is important therefore not to simply consider that some students are gifted *per se*, or even gifted in science *per se*, or even gifted in chemistry (or biology, or physics, etc.) *per se*. In practice many students may have areas of current capability worthy of a label such as 'gifted' (and so likely to benefit from particular curriculum or instruction) whilst in other areas within the subject their current capabilities may be unremarkable. Ideally education not only matches level of demand to capabilities of students in general, but actually looks to give all learners opportunities to make good progress across different learning goals - including more challenging work in those areas where they show high levels of current achievement.

That said, a different argument might be that all learners should be supported to develop across a wide range of education goals, and that where learners show uneven development it is more sensible to focus learning opportunities in areas where that student is *less* advanced. This could be a sensible argument for example when teachers are working with 'twice exceptional' students - those who show exceptional capabilities in some areas, but may also demonstrate specific learning difficulties (Sumida, 2010; Winstanley, 2007). Perhaps a student shows exceptionally high levels of knowledge and logical thought when working orally, but has specific learning difficulties that make producing written work problematic. In such a situation it could make more sense to focus educational support on developing better writing skills given how important these are likely to be to general academic success. Even here, it is questionable whether support to overcome a learning difficultly should preclude opportunities to make good progress in an existing area of strength. In general there is an issue here of the extent to which education should aim to facilitate the development of well-rounded learners with broad areas of capability, rather than seek to identify and nurture exceptional talent. Undoubtedly both are important aims, but in particular cases judgements need to be made about where to focus educational support and where to direct the precious resource of the learner's study time.

Informing such judgements requires evidence-based knowledge. Mozart's prolific talent was nurtured by opportunities to extend himself in music from a young age. In retrospect we might judge the young Mozart's education was unbalanced - but we may not think that he would have been better served (or that the world would be a better place) had the authorities persuaded the young Wolfgang's family that he should spend less time on his music to develop his knowledge and skills across a much wider range of subjects and activities! Darwin was not identified as an exceptional scholar when he was young, but he had an intense fascination with natural history, seemingly at times to the detriment of his studies of the wider curriculum. With the benefit of hindsight it seems clear now that Darwin was a gifted learner in the sciences — but then hindsight is itself said (sarcastically) to be a wonderful gift! Research is needed to better understand the profile of capabilities (and potentialities) brought to science learning by learners in general, and those we might label 'gifted' in particular.

Science Provision Needs to Offer Suitable learning Opportunities to Allow Students (And in particular Here, Gifted Students) to Make Progress in Their Learning

As suggested earlier in this chapter, there are very good reasons to be concerned about the educational provision provided in science education for those considered gifted in science. These students are a resource for society, having particular potential to contribute to future scientific and technological advances in a world facing serious problems (some of which are due to the - historically unprecedented - scale of human populations and the resource implications of the lifestyles people aspire to).

Moreover, there is a moral issue in considering the nature of education for the gifted. If society takes responsibility for the education of its young people, and (as it often does in modern societies) makes education not only an entitlement but also compulsory for those within a certain age range, then that society has responsibilities not to waste the time of the young whilst attending school. Compulsory education must bring an entitlement to provision that supports meaningful and substantive learning. That is, if we require gifted students to be in school, we should ensure school offers them opportunities to fully benefit from being there that make their attendance worth their while. Yet it is suspected that in many educational contexts, gifted students spend much of their time 'treading water' - undertaking 'busy work' which does not engage their creativity or challenge their thinking. Educators should not have the right to waste the time of those who are required to attend our classes.

Research therefore needs to support teachers and curriculum developers to provide gifted learners in science with suitable provision to give them the opportunities to make genuine progress in their learning. There are already many studies that report offering particular inputs for gifted learners (whether within normal classes, in gifted streams, as enrichment in the school outside normal curriculum, or as special programmes attended away from the home school). This research is theorised to different degrees, and evaluated at various levels - some programmes have been subject to extensive systematic evaluation, but this has often not been possible for many small-scale initiatives.

The question for the research programme is what forms of provision do genuinely offer suitable learning opportunities to allow gifted students to make substantive progress in their science learning. The programmatic nature of the research is such that for this question to be effectively tackled, research needs to have already made progress on developing theory around some of the earlier core commitments (so that researchers are able to characterise the students they are working with and their potential for further learning).

Particular Science Provision is Indicated for Those Learners in Any Cohort Who Are Too Advanced to Find Substantive Learning Opportunities in the Standard Provision

Implicit in this last point is that - given some of the ideas already developed - it is not reasonable to assume that standard educational provision - normal curriculum taught in normal classes - offers sufficient scope to challenge the most able learners and support their development and learning progression. The question suggested by this candidate for a research programme hard core commitment is clearly along the lines of *what are the particular kinds of provision which are suitable for gifted science learners, that allow teachers to match instruction to these students' readiness for further learning*. As with the previous point, progress here is dependent upon having a good understanding of the current capabilities of learners and how these can act as the base for further learning that taps into students' potentialities.

This question needs to be explored within wider concerns. To what extent is it desirable to separate out gifted learners from their peers either physically (by teaching them in a different place) or marking them out through setting them clearly different work to others in they class? It is argued that gifted learners need to work with others of similar characteristics to engage in the kinds of peer dialogue that will support learning and engage them in subject matter (Taber, 2007a) - and, sometimes, gifted learners may be - at least to some extent - socially rejected by those who regard them as too nerdy or geeky. Indeed it is sometimes claimed that many gifted learners may deliberately underplay their capabilities to better fit in socially with their classmates. Yet some gifted learners do manage to fit well within heterogenous classes, and setting up separate classrooms for the gifted (even where numbers make this viable) may only encourage a sense of otherness and difference.

Gifted learners will hopefully develop into gifted adults who contribute to society, and often this will mean socially engaging effectively in the workplace and wider community with others who may have very different intellectual characteristics. Too much separation from a diverse range of peers during formative years may make successful transition to adult social contexts more difficult.

Another problem with setting up a gifted stream - or even gifted school - is that it is based on the dichotomous notion that some learners *are* gifted and some are *not*. Much of the presentation that has led to this point in the chapter has argued that there are gifted learners who have quite different characteristics than most of their peers - *in some regards*. Yet it has also been suggested (a) student capabilities tend to lie on broad distributions, so there is not likely to be a clear gulf between groups of learners classed as 'gifted' and 'non-gifted' (even when there is clear gulf between *some* learners who may be classed as 'gifted' and *some* of their peers) and (b) that learners often have an uneven profile of capabilities, even within science, so they may be gifted in the context of some topics or skills areas, but not others.

Moreover, there is a wide suspicion in some education systems of any kind of labelling and segregation of individuals because it can itself bring about effects (Rosenthal & Jacobson, 1968). In the English system of state-funded secondary education set up in the 1940s it was decided to provide different kinds of schools for different kinds of learners. As a result of a test taken one day in the final year of primary school students were (in effect) labelled as academic or not. This system was later abandoned because it was realised that whilst it was not possible to identify academic potential in such a simplistic way, the assignment of a learner to the non-academic school made it extremely unlikely that learner could ever work their way back into the stream being prepared for university entrance examinations or possible entry to the professions - especially when such learners came from less advantaged socioeconomic backgrounds. It was not impossible to switch school type after eleven, but the expectations and different provision in the 'modern' schools made it very unlikely a student there would be found to be working at a level sufficient to be able to be unproblematically moved to the 'grammar' school (Taber, 2012b). Moreover, certainly in the British tradition, the notion of different kinds of schools suited to different kind of learners was widely seen as implying something about the worth of the learner and the work they were being prepared for: grammar > technical > modern.

Some of these issues are avoided when learners remain in the same classes but are set different work. This may also be seen as divisive, however, if a few 'gifted' students are given something different to do than the rest of the class - especially if they are given something perceived as more interesting that 'gets them out of' doing the more routine work assigned to the rest of the class.

This scenario also has the potential for leading to a situation where most of the teacher's time is spent dealing with the default work in the class, with only occasional forays to see how the gifted learners are getting on. We might expect gifted learners to have more advanced metacognition (but not necessarily, given the point about different profiles that students may have) and we might feel that being given more expansive tasks that need to be planned and managed by the learner with less teacher input is a good thing (Shore & Dover, 2004). In general this is likely to be so (Taber, 2007a), but if we adopt Vygotsky's notion of the ZPD (as discussed above) we will also be aware that learners are facilitated to make the most progress when scaffolded by a more experienced person (such as the teacher).

One possible direction for research is exploring opportunites to employ the most advanced students as teaching auxiliaries in classes. This must not be (as in the Victorian schoolhouse model whether older students acted as 'monitors') simply a means of 'using' gifted students as a (cheap) supplementary teaching force, but rather recognising the value to learning of having responsibility for teaching others.

The focus would be on how preparing for and actually teaching others can support the development of a better-integrated and deeper understanding of subject matter for the 'teacher' (Taber, 2009a). This approach allows the development of personal organisation, metacognitive and social educational objectives as well as conceptual learning - but will not suit all gifted learners, and possibly is more feasible in some curriculum topics and educational contexts than others. This is just one suggestion for a line of research that can be developed.

There are then considerations that pull us in different directions - and that is why research is needed to test out these different considerations in various educational contexts to find out what kind of provision best meets the (intellectual, and other) needs of gifted learners in science - or indeed different kinds of gifted learners doing various kinds of science learning. Conceptualising such research within a common overall programme (perhaps along the lines suggested here) can help to build coherence between isolated studies.

Conclusion and Recommendations

The argument made here is that

a) there is a pressing need for the science education community to focus coordinated research attention on the issue of teaching the highest achievers, the gifted, in science;
b) areas of enquiry make progress through the establishment and development of well-defined scientific research programmes (Lakatos, 1970), which clarify core assumptions and establish lines of work consistent with those assumptions; so
c) there is value in setting out the starting points for a research programme exploring educational provision for the gifted in science.

One issue that this characterisation of the field has attempted to highlight is the problematic nature of the label of the gifted learner (in science, or more generally). A progressive research programme has to respond to important issues around the discomfort of many educators in some educational contexts of labelling students with terms such as 'gifted'. The term itself is considered inappropriate by some due to its traditional meaning of having been given a gift by fiat. This suggests people 'just are' gifted (or not) through the accident of who they happen to be, and there is nothing they or anyone else can do to change that. Most educators believe, based in part at least on the evidence of educational practice, that people have considerable potential to develop - and indeed that labelling people may lead to expectations (in teachers and students, in regard to those judged as gifted, and those judged as not gifted) that can in themselves have effects.

The approach taken here is to argue that giftedness/non-giftedness should not be seen as a dichotomy where decisions will always need to be made about 'cut-offs' within a spectrum of attainment/potentiality, and that giftedness is never an absolute but is always a judgement made relative to some specific context. The learner who is not considered gifted this week in the context of learning about human anatomy (because the standard curriculum provision is suitably demanding) might deserve to be considered gifted next week in the context of studying the gas laws (because these seem intuitively obvious to her and the standard scheme of work on this topic presents no challenge to move her thinking and understanding on), and may indeed be considered gifted in two years' time when anatomy comes around again in the spiral curriculum (as in the meantime she has developed a burning desire to enter medicine and has been obsessively reading everything she can find about human anatomy and physiology).

So the notion of giftedness proposed here is a pragmatic and constructive one: in terms of a need for special attention to ensure opportunities are provided that can be genuinely educative for the particular learner when the standard curriculum to be presented to the wider group of learners does not offer a match in terms of the gifted learner's current state of learning and potential for further progress (in the particular subject, topic, or even activity). Whether individual teachers are comfortable with using the term gifted or not, this is an issue of educational equity - all learners in a course or programme of study are entitled to be provided with genuine opportunities for substantive further learning.

One radical take on the characterisation of the research programme offered here is that it is largely not about gifted learners *per se* at all: but rather is about the need for science teaching that better reflects individual differences. All students are different, all come to class with unique resources for learning: they all know somewhat different things, have different levels of understanding of topics, different interests, different sets of 'alternative conceptions', different aspirations, and - in terms of the discussion above - very different profiles of current strengths and weaknesses (and of current potentials for immediate progression) across and within curriculum subjects. The logic here then might suggest that the notion of gifted learners is both distracting and unnecessary. Responding to the general challenge of identifying students' potential for learning and matching this with suitable learning opportunities would solve the issue of the learners who might be be considered gifted in certain learning contexts without needing the label or inviting issues of cleaving the gifted from the non-gifted as an unjustified dichotomisation. Whether such a general programme would risk losing sight of the particular needs of those who are genuinely outliers on distributions of current capability and immediate potential for further learning is a moot point.

Of course it is not assumed that because someone working in a field promotes a particular model for a research programme (such as that described here) that this will necessarily be taken up by others working on the topic. The fundamental commitments of a research programme have to be shared by those working in a field - they have to be committed to! - and they cannot simply be imposed. Perhaps some colleagues would disagree over some of the suggested commitments here - either radically, or at least in detail. This chapter has argued the case for the need for seeing this important area of work in terms of a research programme, and has put forward what seems a reasonable characterisation of the basis of that programme. If others with interests in this area are provoked to disagree over specifics of this proposal, but within a mindset of seeking to establish common ground across researchers, then at least the argument here might initiate fertile conversations about how to best understand this important area of research.

REFERENCES

Adey, P. (1999). *The Science of Thinking, and Science For Thinking: a description of Cognitive Acceleration through Science Education (CASE).* Geneva: International Bureau of Education (UNESCO).

Ausubel, D. P. (2000). *The Acquisition and Retention of Knowledge: a cognitive view.* Dordrecht: Kluwer Academic Publishers.

Bliss, J. (1993). The relevance of Piaget to research into children's conceptions. In P. J. Black & A. M. Lucas (Eds.), *Children's Informal Ideas in Science* (pp. 20-44). London: Routledge.

Bruner, J. S. (1960). *The Process of Education.* New York: Vintage Books.

Charlton, B. G. (2009). Why are modern scientists so dull? How science selects for perseverance and sociability at the expense of intelligence and creativity. *Medical Hypotheses,* 72(3), 237-243. doi: 10.1016/j.mehy.2008.11.020

Claxton, G. (1993). Minitheories: a preliminary model for learning science. In P. J. Black & A. M. Lucas (Eds.), *Children's Informal Ideas in Science* (pp. 45-61). London: Routledge.

Collins, H. (2010). *Tacit and Explicit Knowledge.* Chicago: The University of Chicago Press.

Cropley, A. J., & Dehn, D. (Eds.). (1996*). Fostering the Growth of High Ability: European Perspectives.* Norwood, N. J.: Ablex Publishing Corporation.

Dai, D. Y., Swanson, J. A., & Cheng, H. (2011). *State of Research on Giftedness and Gifted Education: A Survey of Empirical Studies Published During 1998—2010 (April).* Gifted Child Quarterly, 55(2), 126-138. doi: 10.1177/0016986210397831

Darwin, C. (1859/1968). *The Origin of Species by Means of Natural Selection, or the preservation of favoured races in the struggle for life.* Harmondsworth, Middlesex: Penguin.

Department for Education and Skills. (2002). *Teaching able, gifted and talented pupils: overview, 2002.* Retrieved 14th January 2004, from http://www.standards. dfes.gov. uk/ midbins/keystage3/

Diezmann, C. M., & Watters, J. J. (1997). Bright But Bored: Optimising the environment for gifted children. *Australian Journal of Early Childhood,* 22(2), 17-21.

Driver, R., & Easley, J. (1978). Pupils and paradigms: a review of literature related to concept development in adolescent science students. *Studies in Science Education,* 5, 61-84.

Driver, R., & Erickson, G. (1983). Theories-in-action: some theoretical and empirical issues in the study of students' conceptual frameworks in science. *Studies in Science Education,* 10, 37-60.

Duit, R. (2009). *Bibliography - Students' and Teachers' Conceptions and Science Education. Kiel,* Germany: IPN - Leibniz Institute for Science and Mathematics Education.

Fensham, P. J. (2004). *Defining an Identity: The evolution of science education as a field of research.* Dordrecht: Kluwer Academic Publishers.

Fraser, B. J., Tobin, K. G., & MacRobbie, C. J. (Eds.). (2012). *Second International Handbook of Science Education.* Dordrecht: Springer.

Gallagher, J., Harradine, C. C., & Coleman, M. R. (1997). Challenge or boredom? Gifted students' views on their schooling. *Roeper Review,* 19(3), 132-136. doi: 10.1080/02783199709553808.

Gilbert, J. K. (1995). *Studies and Fields: Directions of Research in Science Education. Studies in Science Education,* 25, 173 - 197. doi: 10.1080/03057269508560053.

Gilbert, J. K., & Newberry, M. (2007). The characteristics of the gifted and exceptionally able in science. In K. S. Taber (Ed.), *Science Education for Gifted Learners* (pp. 15-31). London: Routledge

Gilbert, J. K., Osborne, R. J., & Fensham, P. J. (1982). *Children's science and its consequences for teaching. Science Education,* 66(4), 623-633.

Gilbert, J. K., & Swift, D. J. (1985). Towards a Lakatosian analysis of the Piagetian and alternative conceptions research programs. *Science Education,* 69(5), 681-696.

Gilbert, J. K., & Watts, D. M. (1983). Concepts, misconceptions and alternative conceptions: changing perspectives in science education. *Studies in Science Education,* 10(1), 61-98.

Gould, S. J. (1992). The Mismeasure of Man. London: Penguin.

Jenkins, E. W. (2007). School science: a questionable construct? *Journal of Curriculum Studies,* 39(3), 265-282.

Key Stage 3 National Strategy. (2004). *Pedagogy and Practice: Teachng and learning in secondary schools. Leadership guide.* London: Department for Education and Skills.

Kirschner, P. A., Sweller, J., & Clark, R. E. (2006). Why minimal guidance during instruction does not work: An analysis of the failure of constructivist, discovery, problem-based, experiential, and inquiry-based teaching. *Educational Psychologist,* 41(2), 75-86.

Klahr, D. (2009). "To every thing there is a season, and a time to every purpose under the heavens": What about direct instruction? In S. Tobias & T. M. Duffy (Eds.), *Constructivist Instruction: Success or failure*? (pp. 291-310). New York: Routledge.

Kuhn, T. S. (1996). The Structure of Scientific Revolutions (3rd ed.). Chicago: University of Chicago.

Lakatos, I. (1970). Falsification and the methodology of scientific research programmes. In I. Lakatos & A. Musgrove (Eds.), *Criticism and the Growth of Knowledge* (pp. 91-196). Cambridge: Cambridge University Press.

Lakatos, I. (1974/1981). Scence and pseudoscience. In S. Brown, J. Fauvel & R. Finnegan (Eds.), *Conceptions of Enquiry* (pp. 114-121). London: Methuen.

Lederman, N. G., & Abell, S. K. (2014). *Handbook of Research on Science Education:* Volume II. New York: Routledge.

Levinson, R. (2007). Teaching controversial socio-scientific issues to gifted and talented students. In K. S. Taber (Ed.), *Science Education for Gifted Learners* (pp. 128-141). London: Routledge.

Martin, A. J. (2011). Holding back and holding behind: grade retention and students' non-academic and academic outcomes. *British Educational Research Journal,* 37(5), 739-763. doi: 10.1080/01411926.2010.490874

Moursund, D. (2006). *Computers in Education for Talented and Gifted Students: A Book for Elementary and Middle School Teachers. Eugene*, Oregon: University of Oregon.

National Research Council Committee on Scientific Principles for Educational Research. (2002). *Scientific Research in Education.* Washington DC: National Academies Press.

Nickerson, R. S. (1998). Confirmation bias: A ubiquitous phenomenon in many guises. *Review of General Psychology,* 2(2), 175-220.

Osborne, R. J., & Wittrock, M. C. (1983). Learning Science: a generative process. *Science Education,* 67(4), 489-508.

Phillips, N., & Lindsay, G. (2006). Motivation in gifted students. *High Ability Studies,* 17(1), 57-73. doi: 10.1080/13598130600947119.

Piaget, J. (1970/1972). *The Principles of Genetic Epistemology* (W. Mays, Trans.). London: Routledge & Kegan Paul.

Popper, K. R. (1934/1959). *The Logic of Scientific Discovery.* London: Hutchinson.

Popper, K. R. (1974/1992). *Unended Quest: An intellectual autobiography.* London: Routledge.

Popper, K. R. (1989). *Conjectures and Refutations: The Growth of Scientific Knowledge,* (5th ed.). London: Routledge.

Riggs, P. J. (1992). *Whys & Ways of Science: Introducing philosophical and sociological theories of science.* Melbourne: Melbourne University Press.

Riley, T., Bevan-Brown, J., Bicknell, B., Carroll-Lind, J., & Kearney, A. (2004). *The Extent, Nature and Effectiveness of Planned Approaches in New Zealand Schools for Providing for Gifted and Talented Students.* New Zealand: Ministry of Education Research Division.

Rosenthal, R., & Jacobson, L. (1968). *Pygmalion in the classroom: teacher expectation and pupils' intellectual development.* New York: Holt, Rinehart & Winston.

Ryan, J., & Louie, K. (2007). False Dichotomy? 'Western' and 'Confucian' concepts of scholarship and learning. *Educational Philosophy and Theory,* 39(4), 404-417. doi: 10.1111/j.1469-5812.2007.00347.x

Sadler, T. D. (Ed.). (2011). *Socio-scientific Issues in the Classroom: Teaching, learning and research* (Vol. 39). Dordrecht: Springer.

Shayer, M., & Adey, P. (1981). *Towards a Science of Science Teaching: Cognitive development and curriculum demand.* Oxford: Heinemann Educational Books.

Shore, B. M., & Dover, A. C. (2004). Metacognition, intelligence and giftedness. In R. J. Sternberg (Ed.), *Definitions and Conceptions of Giftedness* (pp. 39-45). Thousand Oaks, California: Corwin Press.

Smardon, R. (2009). Sociocultural and cultural-historical frameworks for science education. In W.-M. Roth & K. Tobin (Eds.), *The World of Science Education: Handbook of Research in North America* (pp. 15-25). Rotterdam, The Netherlands: Sense Publishers.

Solomon, J. (1993). Four frames for a field. In P. J. Black & A. M. Lucas (Eds.*), Children's Informal Ideas in Science* (pp. 1-19). London: Routledge.

Solomon, J. (1994). The rise and fall of constructivism. *Studies in Science Education,* 23, 1-19.

Stepanek, J. (1999). *Meeting the Needs of Gifted Students: Differentiating Mathematics and Science Instruction.* Portland, Oregon: Northwest Regional Educational Laboratory.

Sternberg, R. J. (1993). *The concept of 'giftedness': a pentagonal implicit theory The Origins and Development of High Ability* (pp. 5-21). Chichester: John Wiley & Sons.

Sternberg, R. J. (2010). *Assessment of gifted students for identification purposes: New techniques for a new millennium. Learning and Individual Differences,* 20(4), 327-336. doi: 10.1016/j.lindif.2009.08.003

Sumida, M. (2010). Identifying twice-exceptional children and three gifted styles in the Japanese primary science classroom. *International Journal of Science Education,* 15(1), 2097-2111.

Sutherland, P. (1992). *Cognitive Development Today: Piaget and his critics.* London: Paul Chapman Publishing.

Taber, K. S. (1998). An alternative conceptual framework from chemistry education. *International Journal of Science Education,* 20(5), 597-608.

Taber, K. S. (2000). Finding the optimum level of simplification: the case of teaching about heat and temperature. *Physics Education,* 35(5), 320-325.

Taber, K. S. (2006). Beyond Constructivism: the Progressive Research Programme into Learning Science. *Studies in Science Education,* 42, 125-184.

Taber, K. S. (2007a). *Enriching School Science for the Gifted Learner.* London: Gatsby Science Enhancement Programme.

Taber, K. S. (2007b). Science education for gifted learners? In K. S. Taber (Ed.), *Science Education for Gifted Learners* (pp. 1-14). London: Routledge.

Taber, K. S. (2009a). Learning from experience and teaching by example: reflecting upon personal learning experience to inform teaching practice. *Journal of Cambridge Studies,* 4(1), 82-91.

Taber, K. S. (2009b). *Progressing Science Education: Constructing the scientific research programme into the contingent nature of learning science.* Dordrecht: Springer.

Taber, K. S. (2010). Constructivism and Direct Instruction as Competing Instructional Paradigms: An Essay Review of Tobias and Duffy's Constructivist Instruction: Success or Failure? *Education Review,* 13(8), 1-44. http://www.edrev.info/essays/ v13n8index.html

Taber, K. S. (2011). Constructivism as educational theory: Contingency in learning, and optimally guided instruction. In J. Hassaskhah (Ed.), *Educational Theory* (pp. 39-61). New York: Nova. Retrieved from https://camtools.cam.ac.uk/wiki/eclipse/ Constructivism.html.

Taber, K. S. (2012a). *The cultures of science education across the world. Studies in Science Education,* 48(2), 229–235. doi: 10.1080/03057267.2012.738021

Taber, K. S. (2012b). Meeting the needs of gifted science learners in the context of England's system of comprehensive secondary education: the ASCEND project. *Journal of Science Education in Japan,* 36(2), 101-112.

Taber, K. S. (2013a). *Modelling Learners and Learning in Science Education: Developing representations of concepts, conceptual structure and conceptual change to inform teaching and research.* Dordrecht: Springer.

Taber, K. S. (2013b). Three levels of chemistry educational research. *Chemistry Education Research and Practice,* 14(2), 151-155.

Taber, K. S. (2014a). [Essay review of] Second International Handbook of Science *Education. Science & Education.* doi: DOI 10.1007/s11191-014-9725-7

Taber, K. S. (2014b). Methodological issues in science education research: a perspective from the philosophy of science. In M. R. Matthews (Ed.), *International Handbook of Research in History, Philosophy and Science Teaching* (pp. 1839-1893): Springer Netherlands.

Taber, K. S. (2014c). Prior Knowledge. In R. Gunstone (Ed.), *Encyclopedia of Science Education.* Berlin-Heidelberg: Springer-Verlag.

Taber, K. S. (2014d). *Student Thinking and Learning in Science: Perspectives on the nature and development of learners' ideas.* New York: Routledge.

Taber, K. S. (Ed.). (2007c). *Science Education for Gifted Learners.* London: Routledge.

Taber, K. S., & Cole, J. (2010). The CREST awards scheme: Challenging gifted and talented students through creative STEM project work. *School Science Review,* 92(339), 117-126.

The National Strategies. (2008). *Gifted and talented education. Guidance on preventing underachievement: a focus on exceptionally able pupils.* 00066-2008KT-EN. Nottingham: Department for Children, Schools and Families.

Vygotsky, L. S. (1978). *Mind in Society: The development of higher psychological processes.* Cambridge, Massachusetts: Harvard University Press.

West, A. (2007). Practical work for the gifted in science. In K. S. Taber (Ed.), *Science Education for Gifted Learners* (pp. 172-181). London: Routledge.

Winstanley, C. (2007). Gifted science learners with special educational needs. In K. S. Taber (Ed.), *Science Education for Gifted Learners* (pp. 32-44). London: Routledge.

Ziegler, A., & Raul, T. (2000). Myth and Reality: A review of empirical studies on giftedness. *High Ability Studies,* 11(2), 113-136. doi: 10.1080/13598130020001188

In: New Developments in Science Education Research
Editor: Nathan L. Yates
ISBN: 978-1-63463-914-9

Chapter 2

RESEARCH AND INNOVATION ON LEARNING WITH AND ABOUT ADVERTISING IN SCIENCE EDUCATION

***Nadja Belova*[*] *and Ingo Eilks*[†]**
University of Bremen, Department of Biology and Chemistry, Institute of Science Education (IDN) - Didactics of Chemistry, Bremen, Germany

ABSTRACT

In our everyday lives we are surrounded by a huge volume of advertising, which displays a dizzying amount of variety. Modern citizens need an ability to critically analyze this type of media offerings. Advertising is addressed in different school subjects, primarily the humanities and social sciences. In these subjects, the main foci of education which examines advertising include the areas of advertising-specific language, images, illustrations, the use of stereotypes, various strategies of persuasion, etc. But advertising also contains scientific information, which is used either explicitly or implicitly. Scientific claims are made in order to bolster the credibility of ad campaigns and to underline the effectiveness of certain products. Understanding the use (and abuse) of such information is an important skill for critical consumers. However, learning through and about advertising has yet to play any significant role in modern science education up to the present. This chapter describes a project aimed at researching and finding innovative learning approaches to advertising education in the science classroom. It covers both learning through and learning about advertising. The current chapter contains a theoretically justified answer on the topic, including both research on teacher's perceptions of the issue and evidence-based development of the curriculum and corresponding pedagogies. It also addresses the implications for innovating science education with regard to a higher level of critical media literacy.

---

[*] Nadja Belova: University of Bremen, Department of Biology and Chemistry, Institute of Science Education (IDN) - Didactics of Chemistry, Leobener Straße NW2, 28334 Bremen, Germany. Phone: +49 421 218-63287, fax: +49 421 218-63288, email: n.belova@uni-bremen.de.

[†] Prof. Dr. Ingo Eilks: University of Bremen, Department of Biology and Chemistry, Institute of Science Education (IDN) - Didactics of Chemistry, Leobener Straße NW2, 28334 Bremen, Germany. Phone: +49 421 218-63280, fax: +49 421 218-63288, e-mail:ingo.eilks@uni-bremen.de.

INTRODUCTION

The mass media is one of the main channels which citizens use to access information in their everyday lives. Due to the growing role of the mass media, media literacy was a topic addressed internationally in 1946 by UNESCO as an element of fundamental education for everyone (Holmes et al., 1947). The growing role of traditional and digital media in our lives has also influenced formal education during the past few decades. Media literacy was therefore embedded in school curricula all over the world as a response to this need. Several self-evident examples are 'Media studies' in Australia (SACE Board of South Australia, 2014), 'Journalism' in the US (Hall and Aimone, 2008), and 'Media education (Medienbildung) across disciplines' in Germany (Landesinstitut für Schule Bremen, 2012).

In nearly all types of media, advertising has played a growing role (Hartnett, 2006). In addition to traditional methods of advertising (newspapers, magazines, radio and television), consumers are now exposed to advertisements at a far greater level than ever before. Advertising permeates their lives and includes such newer media as the Internet and social and professional networking sites. Consumers are forced to adapt in order to process and filter the vast surge of advertising information with which they are being bombarded by a multitude of sources. Advertising has a huge impact on children's and young adults' lifestyles and behavior, for example, their food consumption (Villani, 2001).

In particular, the group of media-geared young people belongs to the largest number of persons who come into increasing levels of contact with advertising. Because of this, education needs to contribute to critical consumer behavior by developing multidimensional media literacy in citizens. Part of this education will have to address people's ability to critically evaluate and responsibly react to all forms of advertising (Brucks, Armstrong and Goldberg, 1988). The required skills which need to be developed include the abilities to both critically evaluate hidden messages and facts in advertisements and to explicitly understand the mechanisms used to create advertisements and subliminally influence uninformed consumers (UNESCO, 2011).

A strong link to science and technology exists in a broad range of advertised products presented in both local and international markets (McSharry and Jones, 2002). Therefore, we must ask the question: Should science education contribute to the development of critical media literacy in our students when they are confronted with science- and technologically-based advertisements?

The claim that science education should more thoroughly contribute general educational skills is not a new one. In Sjöström's understanding of *Bildung*-oriented science education (Sjöström, 2013), science lessons should seek to promote the development of general educational skills via contention with topics selected from science and technology. This will aid in preparing the younger generation in becoming responsible, self-directed citizens in society (Hofstein, Eilks and Bybee, 2011; Holbrook and Rannikmäe, 2007).

Within this framework of more societally-oriented, skills-promoting science education, one sub-set of innovations is further based upon the use of socio-scientific issues (SSI) in the science classroom (Sadler, 2004; 2011).

SSI-based science education serves as a motivating context to promote meaningful science learning. Simultaneously it becomes a catalyst which promotes general educational skills, especially argumentation and decision-making skills (Albe, 2008).

Within the SSI movement the socio-critical and problem-oriented approach to science education was developed in Germany (Marks and Eilks, 2009; Marks, Stuckey, Belova and Eilks, 2014). This approach attempts to construct a consistent model in which to operate SSI-based science teaching. It was developed into a five-step model for each of the curriculum units (Marks and Eilks, 2009). The introduction to a topic is performed using authentic media artifacts, e.g. newspaper articles, brochures printed by pressure groups, reports broadcast by TV stations or (in our case) advertising. The topics selected must allow for real decisions to be negotiated by the learners. Issues are inappropriate if they allow only one-sided solutions or when they cannot be openly debated due to scientific, ethical, or sociological reasons. The activities performed within the lesson plan challenge students to make up their own minds and to verbalize their opinions on the topic in an open forum. Such conditions allow the expression of one's personal point-of-view without the individual being judged, censored or condemned as an outsider by the rest of the group. Reflection upon how society handles SSI is carried out by mimicking an authentic societal practice dealing with various kinds of questions. Different pedagogies have so far been developed for several topics, which allow the contrasting of societal perspectives on a given topic and the societal decision-making processes taking place. These pedagogies include not only conventional methods like role-playing and business games (Feierabend and Eilks, 2011), but also more innovative pedagogies such as working as a journalist (Marks, Otten and Eilks, 2010), a professional product tester (Burmeister and Eilks, 2012) or an advertising specialist (Stuckey, Lippel and Eilks, 2012).

SSI-based science education about advertising is special because in addition to being used a as medium for learning ('learning with advertising'), advertisements can also become the SSI itself ('learning about advertising') with the help of the socio-critical and problem-oriented teaching model mentioned above.

Not only can we illustrate certain scientific topics through appropriate advertising, we can also discuss the scientific information used in advertising, how it is presented, what effect it has on the credibility of the ad, whether incorrect and/or misleading information is used, etc. Advertising was mentioned as early as 1938 by Cox, McCollum and Watkins as a fruitful topic for science lessons. Despite this fact, however, addressing critical media literacy in science education is still rare, regardless of whether you are speaking about learning with advertising or about advertising.

This chapter outlines a project on the implementation of advertising in science education which aims to bridge this gap. The project includes setting a theoretical framework in order to justify the use of advertising in the science classroom. It also covers research on both teachers' and students' perceptions of the topic and it develops the necessary curriculum and its corresponding pedagogies.

THEORETICAL BACKGROUND

Media literacy found its way into curricula and classrooms in the US and UK around the middle of the 20th century (Holmes, et al., 1947). This was mainly done to provide students with skills to protect themselves from the influence of the fast-growing media world (Hobbs and Jensen, 2009). In 1982, nineteen nations signed the 'Grunwald Document' at the

UNESCO International Symposium on Media Education, in which they officially acknowledged a responsibility to help young people to live in a media-rich world by providing comprehensive media education programs (UNESCO, 1982).

Different definitions and concepts of media literacy can be traced in the literature. Two main dimensions of media literacy were outlined by the UNESCO (2006) as 'reading' and 'writing' media. On the one hand students need to understand the different forms of communication used in the media. On the other they have to be able to create their own media. The European Commission (2007) defines media literacy as the ability to access media, understand them, and critically judge media content, including the creation of one's own media products. According to Scheibe and Rogow (2012), media literacy is described as an individual having a number of capabilities which enable them to navigate their way through a media-dominated environment. Four concrete goals of media literacy are accessing, analyzing, evaluating and creating media (Hobbs, 2003). In line with the growing importance of media literacy in general, science in the media currently became also a field in science education research. However, media research in science education still remains limited and is always only focused on specific media types, most generally media such as newspapers or magazines. Examples from the past include Hopkins (1925), Martin (1945), Koelsche (1965), Brossard and Shanahan (2006), and Rundgren, Chang-Rundgren, Tseng, Lin and Chang (2012). An extensive review on the research on news media in the science classroom was recently published by McClune and Jarman (2012).

The objectives for strengthening learning about science in the media include linking classrooms to the outside world and making science lessons more relevant to the students. The corresponding skills encompass being able to critically evaluate media offers and gaining competence in dealing with understanding socio-scientific issues debated in both the news media and the public arena (McClune and Jarman, 2012; Stuckey et al., 2013). In summary, dealing with science-related news and being able to logically and critically discuss such items is viewed as a crucial requirement of becoming a modern citizen (e.g. Elliott, 2006; Marks et al., 2010; McClune and Jarman, 2012; Rundgren et al., 2012; Stuckey et al., 2013).

Another important aspect of learning about the use of science in the media is teaching students about how media products are made and how scientific knowledge is incorporated into them (Eilks, Nielsen and Hofstein, 2014). Science writers for the public primarily follow the basic principles of journalism; scientific accuracy often lags behind (Hansen, 1994).

Thus, a problem arises whenever newsworthiness is placed before science (Priest, 1999; Eilks et al., 2014). In the case of advertising, Belova and Eilks (2014) recently provided a model of filtered information which illustrates how scientific information is selected and transformed on its way to publication in a media source (see also Eilks et al., 2014). This model provides a helpful justification for implementing learning about the media in science education. Klosterman, Sadler and Brown (2012) also point out: 'We see important connections between scientific and media literacy, especially in today's media-rich environments' (p. 62). In consequence, Chang, Rundgren and Rundgren (2014) suggest a concept of 'scientific media literacy' combining scientific literacy with media literacy as an essential part of relevant science education.

In different countries, the national media show widely varying amounts of scientific programming. This situation can lead to different perceptions of science, especially among children and adolescents.

Research has revealed that the media affect pupils' perceptions of both science and its nature. Dhingra (2003) suggests that television-mediated science is a 'significant force' (p. 234) and that it contributes to influencing the views of students on the nature of science. She points out that different programs (documentary, magazine-format programming, network news, dramatic or fictional programming) provide students with different pictures of the nature of science (ranging from 'always valid, final-form science' to 'scientific findings are open to interpretation'). Since students see the use of media in science classrooms as very motivating and memorable, Dinghra concludes that schools cannot ignore the powerful effects of media and other outside forces.

Nisbet et al. (2002) indicated that different media types affect personal perceptions differently with respect to media influence on the scientific views of consumers. They found that heavy television consumption tends to give viewers only limited knowledge about science, while newspapers seem to lead to greater expertise. They concluded that television very often presents a negative image of science and scientists: the image of the 'strange and scary scientist' (p. 603). However, it was also shown that scientific knowledge leads to a more critical, skeptical view of media information. After a health literacy intervention program, for instance, Hove, Paek and Isaacson (2011) discovered that middle school students viewed commercial health web sites as much less reliable a source of information than before. It seems that scientifically literate people are more skeptical about the world. Johnson and Pigliucci (2004) made a contrary claim, indicating that possessing a high degree of scientific knowledge does not automatically increase skepticism towards pseudo-scientific information; however, the information selected for their study was highly esoteric.

Beginning with media in general before moving more thoroughly towards advertising, UNESCO defined 'advertising literacy' as a part of media literacy (UNESCO, 2011). It also provided different modules relating to advertising literacy (UNESCO 2006, 2011). The corresponding modules on advertising literacy focus on the role of commercial forces in the media landscape, advertising placement and regulations, political advertising, the evaluation of ad claims, advertising design, and the semiotics of advertising. In these approaches, the content behind advertising does not play any significant role. In the extensive UNESCO curricula, key-words like 'misleading claims' or 'supporting facts' for advertising do appear, but the issue is never addressed in any detail. Few other teaching ideas can be found in the literature, especially in US teacher journals (e.g. Burrows, 1997; Hanuscin, 2002). These ideas mostly focus on using advertising for motivational purposes or to contextualize content knowledge ('learning *with* advertising'). Aspects of deeper evaluation as well as creation of media ('learning *about* advertising') hardly ever appear.

As already mentioned, advertising is an authentic context for science and technology education, which at the same time often is related to complex and controversial scientific information. Research from the European context shows how often science actually appears in advertising campaigns. McSharry and Jones (2002) examined 1,026 advertisements from UK television, 65% of which were science-based, however only 26% of these science-based advertisements were recognized by the consumers as being grounded in science. Pitrelli, Manzoli and Montolli (2006) conducted a similar investigation in Italian newspapers and disclosed that out of 6,542 advertisings, 1,018 contain a scientific reference (15.6%). The lower numbers of hits generated were affected by the selection criteria because this study only counted ads that explicitly used the terms 'science' or 'scientific'.

Children and adolescents (and therefore school students) are an important target group in marketing campaigns. Products are often created especially for them, such as toys, special foods and cosmetics. For this reason a large amount of advertising for these particular product groups is aired on television and distributed via other types of media every day (e.g. Byrd-Bredbenner, 2002; Hang and Auty, 2011). Young people are also viewed as a potential market due to their status as future buyers and consumers (Connell, Brucks and Nielsen, 2014). Consumer research shows that even advertising aimed at adults has many effects on the younger generation. For instance, after showing pre-school children advertisements for different products like jewelry items and spectacles, Constandinidou-Semoglou (2007) found that they were highly attracted by the images and found them entertaining. This was despite the fact that the children evidenced a broad lack of understanding concerning the ads. Preston (2006) carried out similar research with children between the ages of seven and eleven. He found that children consume and understand more adult advertising as they get older; however, they still lack both an in-depth understanding of and interest in 'old people' aspects of life like insurance agencies. Preston concluded by pointing out that 'advertising that has been designed for an adult audience represents a continuum of interest and significance for children' (p. 82).

Deeper insights into children's' understanding of advertising mechanisms can be found in fundamental consumer research studies. This research shows that children normally learn to distinguish advertising from regular programming by the age of five, with three-year-olds already possessing the ability to do so in many cases (e.g. Palmer and McDowell, 1979; Butter, Popovich, Stackhouse and Garner, 1981). Understanding of the intentions and interest groups behind advertising develops by roughly the age of seven (Robertson and Rossiter, 1974). At this age children also begin to realize that advertising does not always tell the truth. The awareness of this fact further intensifies during the teenage years (Ward, 1972; Bever, Smith, Bengen and Johnson, 1975). However, the question of whether pure knowledge of advertising mechanisms can or will protect children from the effects of advertising is still hotly debated in consumer socialization research. Some researchers suggest that the existence of such knowledge is sufficient to support a critical and skeptical approach towards advertising (Friestad and Wright, 1994; Livingstone and Helsper, 2006), while other researchers maintain the view that pure knowledge does not automatically lead to a more critical viewpoint (Rozendaal, Buijzen and Valkenburg, 2009).

Generally, research examining exactly how children use their knowledge of advertising mechanisms is quite spotty. Those studies which are available (Brucks, Armstrong and Goldberg, 1988; Rozendaal, Buijzen and Valkenburg, 2012) clearly show that children between seven and eleven only activate and apply their knowledge when watching advertisements if they have previously been triggered to do so, for example by an educational film on the nature of advertising. Rozendaal, Buijzen and Valkenburg (2012) showed that teaching and learning strategies which make students more aware of their own critical thoughts on advertising (such as think-aloud activities) clearly reduce their advertising susceptibility. Research on new types of advertising shows similar tendencies (An, Jin and Park, 2014).

Children also understand non-traditional advertising methods with much lower levels of personal sophistication as compared to their comprehension of ads by traditional advertising methods (Owen, Lewis, Auty and Buijzen., 2013).

What can science educators learn from the literature, both in the areas of science education publications and other domains such as consumer research? Advertising often contains scientific information or is related to it, but students lack the necessary knowledge to recognize, understand and evaluate it in many cases. Advertising research suggests that activities intensively dealing with advertising and thus increasing student awareness of the corresponding knowledge can lead to a more critical view on advertising. However, those few teaching ideas available from science education rarely focus on critical media literacy goals of evaluating advertising its sources. Although science education seems to be the only place to bridge the gap between meta-knowledge about ads and the science content knowledge employed by them, a connection in the science classroom is seldom brought into being.

Teachers' Views on Advertising in Science Education

In the UNESCO media literacy curriculum (UNESCO, 2011) teachers are addressed as 'principal agents of change' when it comes to 'making full use of their [citizens'] rights to freedom of expression and the right to information' (p. 11). Teachers are the key factor for any success of educational reform (Hattie, 2009). Effective educational reform will only occur when teachers' *a priori* knowledge, attitudes and beliefs are seriously taken into account (Haney, Czerniak and Lumpe, 1996). Therefore, exploring teachers' prior knowledge and addressing their personal beliefs is a necessary first step, if any attempt to change in teaching practices is to be planned (Van Driel, Bulte and Verloop, 2007).

But do teachers use media in their science lessons? They do, but they seem to stick mainly to traditional 'static print materials' (Klosterman, Sadler and Brown, 2012, p. 71) such as newspapers or books. A review by McClune and Jarman (2012) on learning science in the news also detected teachers' preference for print-based, traditional media. Studies that systematically explore how teachers use news in their classrooms are still rare. To date, three studies with such a systematic approach are available (Jarman and McClune, 2001, 2002; Kachan, Guilbert and Bisanz, 2006). One of the most interesting findings in all of the studies is the focus on a limited number of goals. Teachers mainly use news media to create a link between school science and the real world as well as to motivate the students. Other goals such as teaching about the sociological nature of science, or addressing *how* science is presented in the news, and thus promoting media awareness as well as fostering critical media skills, have not been pursued by the teachers in these studies. A lack of preparedness for fully integrating the potential of critical media education via science education is a major issue due to insufficient teacher training (Lauri, Borg, Günnel and Gillum, 2010).

There is a substantial lack of knowledge about teachers' views on the use of advertising in the science classroom. Therefore we conducted a 'discovery' interview study (see Biddle and Anderson, 1986) to explore teachers' viewpoints. The study is based on semi-structured interviews. The interview guide encompasses questions on the teachers' experience with the use of advertising in the science classroom. It also asks for potential ideas on how to include advertising in science teaching and asks questions about the teachers' perception of the topic's relevance. Each section is designated to a general guiding question which is then divided into sub-questions. Additionally, illustrative material was prepared (Belova and Eilks, 2014). Based on the materials, the teachers were asked for spontaneous comments on various pedagogies and teaching materials for integrating advertising with science education in the last section of the interviews.

The sample consisted of twelve teachers. Since the size of the sample was rather small, we tried to make it as heterogeneous as possible in the sense of choosing a meaningful sample group. The teachers represented various age groups and school types. They came from two German federal states (one urban, one more rural) and the gender distribution was almost equal. The teachers from the federal state of Bremen came from the newly implemented secondary comprehensive school 'Oberschule' with very heterogeneous classes, which encompass a broad variety of nationalities, backgrounds and levels of performance. The teachers from Lower Saxony taught in rural areas either in grammar schools (a German school type for higher-achieving students) or in comprehensive schools. In Germany, every teacher has to study and teach at least two subjects. We selected teachers with different combinations, however, one had to be a science subject. The interviews lasted between 25 and 45 minutes and were first audio-taped and then transcribed. The analysis took place following the principles of Qualitative Content Analysis (Mayring, 2000).

Regarding the findings, most of the teachers claimed to have used advertising in their science lessons, usually as an introduction to a certain topic or as a contextualization for experiments. For instance, several teachers used advertising to contextualize experiments on household cleaners, sugar-rich foods, different types of drinks and cosmetic products. None of the teachers used advertising regularly and systematically. The implementation often seemed to fall in line with the teacher's personal media consumption. It was noticeable that many teachers named newspapers as their primarily source of advertisements for school implementation.

Most examples stemmed from chemistry lessons; biology came in second place and only one example came from physics. While the examples from actual implementation of ads came from a very limited number of domains, the ideas that the teachers had while talking about the topic varied to a much larger extent. Such ideas included the implementation of advertising for particular foods and drinks (e.g., yoghurts with levo-rotary lactic acid, products with added vitamins and minerals, protein supplements, hidden sugars), car advertising to contextualize concepts such as acceleration in the physics classroom, pharmaceutical products, the energy industry (renewable energies, house isolation, fuels) as well as general issues critical of blind consumerism.

The latter included ubiquitous advertising for mobile phones, which leads to frequent purchases of new devices and to problems with recycling the old ones. This range of ideas reveals the tendency that teachers generally do have ideas about how to implement advertising into science lessons.

However, they rarely act accordingly in their personal practices. Concerning the actual implementation and ideas, teachers with two science subjects surprisingly presented more examples than their single-science subject colleagues.

When it came to the perception of the general relevance of advertising in school, there was unanimous agreement that students should learn about this type of medium within the framework of a formal education. Educational goals that were suggested included a critical consideration of advertising (due to its manipulative character), students forming a personal opinion, and the promotion of evaluation and decision-making skills in learners. To put it succinctly: *'The kids don't have to believe everything'*. Four teachers explicitly pointed out the potential of science education regarding advertising: *'Especially the sciences provide a totally different view on advertising; they help to scrutinize certain attitudes and to really understand what is behind an ad'.*

Generally, advertising was found to be relevant as a medium for science education due to its high potential for motivating students, its large presence in their lives, and its sizable impact on young people. All of the teachers seemed to be aware of these factors.

Limiting factors of the use of advertising in the science classroom included the lack of time, an overloaded curriculum, advertising being too far removed from subject matter learning and thus not fitting into the regular pathways of science teaching and the accompanying assessment. Other factors included a lack of ideas, the lack of corresponding teaching materials, a fear of losing one's neutrality when talking about brand names, and a lack of preparedness, which was linked to missing personal expertise on advertising. When teachers were shown illustrative materials with advertising-based methods which had been developed in this project, they generally received positive feedback, especially among the younger teachers.

By further analysis, two types of teachers were identified in the sample. The first group, consisting of eight teachers, showed a very open-minded attitude towards implementing ad-based teaching and learning in the science classroom. These teachers found (scientific) media literacy to be a crucial issue. They supported claims that advertising might be a good way to show their students both how science is portrayed in this type of media and how it presents itself in their everyday lives. They also support the idea that science learning in school is far too often distant from the lives of students and therefore perceived irrelevant. Adverting was seen as a motivational way to bridge this gap and make students more literate members of society. All but one teacher with such views came from urban comprehensive schools and most were between 25 and 45 years of age. Four were male and four female.

The second group of teachers can be described as highly skeptical towards integrating teaching and learning about advertising in science. All of these teachers were over 45 years old, taught only science subjects and mainly came from rural grammar schools. These teachers stated that integrating such goals and their corresponding pedagogies would make science lessons deviate too much from content-matter learning. Reasons given in support of this viewpoint were lack of time, non-flexible curricula, lack of available ideas for topics, lack of personal interest, and a fear of losing the structure if their science lessons based on a structure-of-the-discipline curriculum approach. This group consisted of four teachers, three of whom were male.

Scenarios and Pedagogies for Learning with and about Advertising in Science Education

So far, very few science education papers have dealt with the aspects of advertising. In the case of Germany, advertising has only been suggested to introduce certain topics, act as a data-provider for tasks, or serve as an inducement for experiments in which claims must be experimentally tested (Belova and Eilks, 2014). As mentioned above, teaching ideas published internationally mainly reveal the same approaches (e.g. Burrows, 1997). Only very few ideas go further. For example, Stuckey et al. (2012) described chemistry lessons on food sweeteners. The students were supposed to create an ad by themselves and reflect upon the role of the scientific information involved (also with the respect to a certain target group).

Based on the existing literature, Belova and Eilks (2014) recently suggested four potential scenarios for using advertising in the science classroom. The scenarios differ in their complexity and contribution to media literacy:

1 Motivation and illustration: Advertising is used as an illustrative opening for a new topic, with ads being mainly used for motivational purposes. Advertising can be used to introduce a certain term or idea and to motivate the students to think about it. The ad itself is not further discussed in the lessons, since the lessons take another focus.
2 Contextualization of scientific inquiries and tasks: In this case, advertising is used to contextualize scientific inquiries and learning. Advertisements are used to provoke science-related questions and inquiries, which often result in experiments. The students are asked to formulate questions on the reliability of the given ad claims, then develop scientific inquiries to answer them. Advertising can be also used as a 'data provider' for calculations in scientific and technological contexts. In both scenarios advertising is suggested for contextualization and also motivational purposes; the nature of advertising itself is not further discussed.
3 Reflection on the role of how science-related information appears in advertising: This approach suggests an analysis of how science- and technology-related factual content appears in ads. The reflection concerns the intentions behind science-related information in advertising to support claims. It also contemplates how ads may contribute to misleading or suggestive advertising by use of truncated, falsified or even false scientific information, e.g. advertising in which health claims are not sufficiently supported.
4 Meta-cognition on the interplay between science and advertising: In this approach a thorough socio-scientific, skills-oriented perspective is suggested. This perspective brings the interaction of science and technology with advertising into focus, thereby questioning the transfer of information about science and technology into advertising. The difference that makes this role most relevant for critical media literacy is that it explicitly addresses advertising itself, the development of ads, and the principles behind them while linking them directly to the scientific content.

In most of the publications available for science education we see scenarios 1 and 2. Only few papers focus on scenarios 3 and 4 emphasizing learning *'about'* advertising. Stuckey et al (2013) and Belova and Eilks (2014) suggested corresponding scenarios that are based on the following pedagogies:

Analyzing promotional texts: Promotional texts for certain products related to science and technology can be found on websites and in print advertisements. In these texts products are described and their advantages are presented. Such texts often contain scientific information which is used to make the product's advantages more credible. We suggest analyzing such texts from the scientific perspective. Figure 1 shows an excerpt of an original text on a new body lotion (translated from nivea.de, 04.11.2014). The students are first asked to rate the text's credibility, attractiveness and scientific nature. Then they have to mark the scientific terms in the text and search for the meanings and backgrounds of certain terms. Such activities are followed by a discussion of whether or not a marketer in the cosmetics industry must also be a chemistry expert. Finally, the text's credibility, attractiveness and scientific nature are rated again.

> **What makes the In-Shower body lotion different from a normal body lotion or shower product?**
>
> In the In-Shower body lotion no conventional emulsifiers were used, as these can reduce lipid levels during rinsing the product off. Therefore, a unique polymer combination was developed to support the stability of the hydrodispersion. The filler Aluminum Starch Octenylsuccinate repels water and completes the formula of the In-Shower body lotion. In addition, the In-Shower body lotion contains the skin-moisturizing ingredient glycerin.

Figure 1. Excerpt from a promotional text suggested for analysis in the science classroom.

Product / Criterion	A:	B:	C:
Scientific aspects			
Scientific terms in the advert			
Scientific images/charts/graphs in the advert			
Study results in the advert ("scientifically proven")			
Survey results in the advert ("9 out of 10 women…")			
Scientific inventory (flasks, test tubes) in the advert			
Emphasis on certain ingredients (e.g. vitamins)			
Guarantees are given (money-back-guarantee, satisfaction guarantee)			

Figure 2. Excerpt from a grid for rating scientific aspects in advertising.

Advertisement rating grid: To help students become aware of science-related aspects in advertising a grid can be used, which helps to rate certain aspects and to compare different product presentations to one another. Figure 2 shows selected criteria from such a grid. Such criteria include the presence of scientific terms, images, or study results. The grid can also be used to rate other pupils' self-created advertisements. The grid also helps to foster a discussion on the role of science in the marketing of certain product groups (e.g. cosmetics, foods or pharmaceuticals).

Creating advertising: In relation to the media literacy goal of 'creating media,' students are encouraged to create their own advertisements. The students receive selected information on a product which can be techno-scientific, economic, etc. The first step is to sort it into groups of positive and negative information. Based on analysis of the available information the students create advertisings by working in small groups. They have to select the best set of positive information which seems most promising to convince a certain target group about a product. During the presentation of their ads the students must reflect upon which of the ads presented were the most convincing and why. The discussion includes whether employing techno-scientific information is reasonable or not when tailoring an ad for a specific target group. It is also asked if the case might be different for products and/or other target groups. An example is shown in Figure 3.

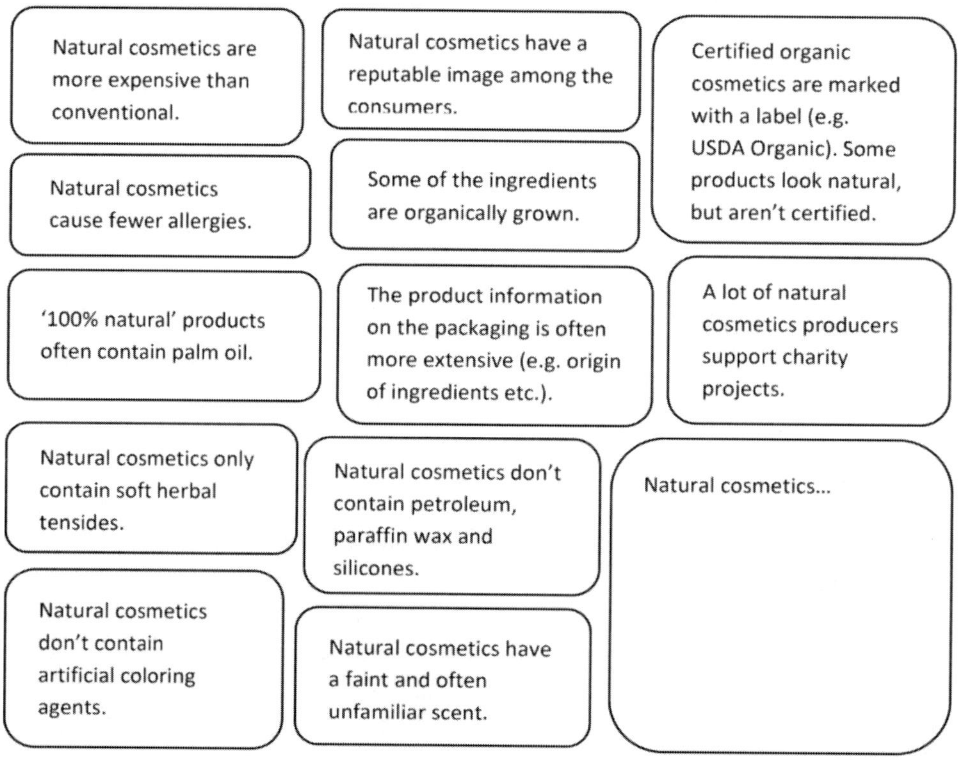

Figure 3. Possible pre-selected promotional information about natural cosmetics. The students evaluate the statements for their suitability for advertising purposes. First, negative or potentially negative statements must be sorted out. Of the positive (or seemingly positive) statements a selection must be made with regard to a target group.

Reflecting on slogans: In this method the students receive a selection of real slogans for a certain product. The students evaluate the slogans using three scales: credibility, attractiveness and the relationship to relevant scientific information. The students realize how different their perceptions of different advertising slogans are. Discussion is initiated which evaluates slogans based on scientific facts or addressing scientific thinking. Pupils also question the role which science-based information plays, how such info is selected and displayed, and whether it affects the credibility and/or attractiveness of advertising in a positive or negative way (Figure 4).

From scientific information to advertising: Another possible teaching scenario reconstructs the way in which an advertisement is created. Students receive a scientific text which provides the basis for a specific product group (e.g. on proteins in the context of protein shakes). Based on this text the students have to develop a concept for a fictive product in small groups. They must take into account issues such as the necessary raw materials, product properties, the selected target group, a possible price range, etc. They have to decide which information from the original text they should use for the product concept they have chosen. As the next step, they have to develop an advertisement for their product. The advertisements are then presented to the other groups and compared to each other. It can be discussed how the scientific information from the original text has been filtered and altered, and how differently the same scientific information can be presented in various advertising campaigns.

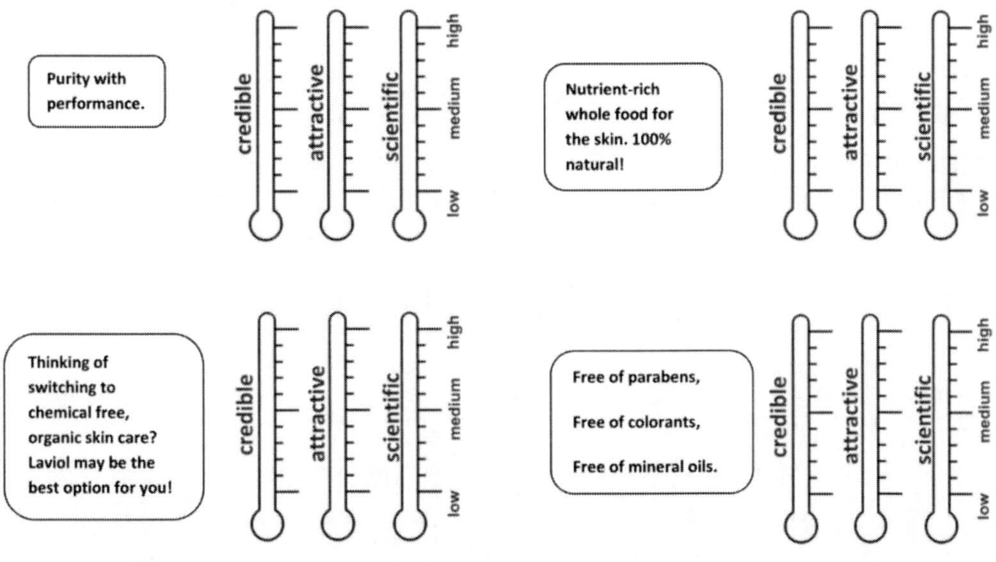

Figure 4. Advertising slogans for natural cosmetics (excerpt from a corresponding worksheet). The students evaluate the credibility, attractiveness and science-relatedness of advertising slogans.

STUDENTS' PERCEPTIONS ON ADVERTISING IN THE SCIENCE CLASSROOM

Just like the situation with teacher viewpoints mentioned above, the available research on students' opinions about using media in the science classroom remains quite spotty. Ten years ago, Lemke (2004) had already noticed that 'there is very little systematic research on students' reactions to science as portrayed in various popular media' (p. 43). The news media are also prevalent when it comes to students forming their perspectives (McClune and Jarman, 2012). Research shows that pupils have limited capabilities when it comes to interpreting science-based news reports (e.g. Kolstø, 2001).

Nevertheless, more recent studies have shown that young people are not fully naïve while dealing with media offerings and are able to critically interpret scientific content (Christensen, 2011). These abilities depend on the circumstances and the nature of the media text. Such positive tendencies might be interpreted as the success of better implemented media in science education programs (McClune and Jarman, 2012). Other findings show that students find science-related news highly attractive and more interesting than textbooks (Halkia and Mantzouridis, 2005; Jarman and McClune, 2002).

In their review, McClune and Jarman summarize these findings by pointing out that 'students are […] attracted by the relevance of science news in the world around them – it is important to see the usefulness of science' (p. 26). Because consumer research shows that advertising widely targets these students as a group (see above), we can see why advertisement-based environments would be regarded as interesting and attractive by them.

Teaching modules focusing on advertising in science education were developed in order to get first insights into how students react to advertisement-based learning environments.

Table 1. Structure of an advertisement-based module on natural cosmetics

Phase and activity	Advertising type scenario #
Investigation of prior knowledge: The students rate the credibility of different claims such as 'Natural cosmetics do not contain any chemicals'. Such claims are most frequently used for advertising.	3
Reflection on advertising slogans: The students get a worksheet with 10 authentic slogans for which they must rate natural cosmetic products regarding their attractiveness, scientific background and credibility. Possible correlations between these three dimensions are also discussed. After the discussion the work sheets are collected by the teacher.	3,4
Introduction: The students are shown a collage with different labels and ads for natural cosmetics which are discussed (also in relation to the previously shown claims).	1, 3
Subject learning: Worksheet on the components of a basic skin cream.	
Jigsaw exercise: The students act as employees of the German national institute of risk assessment and are supposed to formulate their recommendations for different controversial cosmetics ingredients. Beforehand, they have been given information texts on these ingredients which contain positive and negative information.	
Experiment: In a guided inquiry-based scenario, the pupils prepare their own skin cream and decide whether or not to use various ingredients.	2
Creating advertising: The students receive a summary of positive and negative information on natural cosmetics. They have to select the information they consider important and create an ad for their product. In the end they present their ads and discuss whether they have actually used scientific information or not.	4
Final discussion: The students are given back their completed worksheets with the slogan ratings from the beginning of the unit. They discuss whether their opinions and assessments have changed during the module. The claims from the beginning exercise are rated again to see how perceptions have changed.	3,4

These units encompassed several of the methods and strategies described in the previous section. Modules employed the topics of natural cosmetics, protein shakes and biodegradable plastics. There were versions for both junior and senior high school. In a case study one of the modules was used in five learning groups in a northern German middle school (grades 7, 8, 9, 11). Two of the classes were eighth-graders and the total sample consisted of 102 students. The structure of the module is shown in Table 1. Lessons consist of four 45-minute classroom periods. The versions for junior and senior high school differ mainly in the lengths of the texts covering the scientific basics.

Both classroom observations and a questionnaire containing five open-ended questions and ten Likert items were used for evaluation. The open-ended questions were evaluated according to Mayring's (2000) Qualitative Content Analysis (QCA). The first open-ended question asked pupils about the most substantial issues which they had been taught during the module.

Among other things we wanted to see whether the aspects the students remembered stemmed from the domain of scientific content or if they also would name advertising issues. We found that 65% of the statements were solely science-based (*'I have learned a lot about the different ingredients in skin creams and shampoos.'*). A further 15% covered both subject matter and advertising. The final 20% were exclusively related to advertising (*'Advertising sometimes presents scientific information in a confusing way and I have to be careful not to fall for wrong things'*). The participants seemed to be more aware of their personal learning outcomes concerning subject matter and not so much concerning the 'non-traditional' issues such as science-based advertising strategies.

A further two questions asked for feedback on the module (What did you particularly like/not like?). The most frequently mentioned aspects were the hands-on design and preparation of skin cream as well as the student-centered work in general. Only 15% of the pupils mentioned the advertising-based activities among the things they had particularly liked. All of these students came from grades 7 and 8. Very few aspects were regarded as negative by the participants. The most frequent criticisms were a noisy classroom atmosphere during the group work phases and a lack of personal interest in the content.

Fifty-four percent of the students' statements indicated a belief that their decisions might be affected. The main reasons given were: 1) the knowledge gained in the unit would help in evaluating the truthfulness of advertising claims in everyday life and 2) it would now be easier to connect this knowledge to everyday situations. Twenty-eight percent of the students stated that they would not change their purchasing habits, mostly because they will still be influenced by other aspects (such as smell, packaging design, etc.).

The final question asked participants whether advertising should be a recurrent topic in the science classroom. This was answered positively by a great majority of the students (83%). The main reasons for this were the proximity to everyday life issues and the fact that advertising aspects can often be connected to school subject matter. Selected statements include: *'It is good to know what some slogans really mean.'*, *'Only a scientific basis makes it possible to talk about advertising in a serious way.'*, *'This topic is a sort of practical test for school chemistry.'*. Still, some of the older students did not regard this issue as a necessary part of their science lessons. They find content knowledge as being more important. One student stated that in his opinion *'science is about 'bigger' things, about nature and so on, and not about this [advertising]'*. Five students explicitly stated that advertising should remain a part of other subjects such as German.

In the Likert items, the students had to rate the attractiveness of the learning environment as well as their personal interest in advertising in science education. Five items focused on personal consumer behavior. Findings from selected items are presented in Figure 5.

These results show that learning about advertising is generally regarded positively, which is in line with the above-mentioned findings on the news media. We can also see that most students prefer to discuss topics which are not purely science. Most of the students stated that such lessons affected their opinions on cosmetics in particular and advertising in general, however, these two items showed the largest levels of disagreement among the participants.

Classroom observations showed that - among other things - students possess a large amount of knowledge about advertising types and strategies. Most probably this comes from other, non-science subjects. However, most learners have difficulties if they are asked to evaluate advertisements from a scientific perspective.

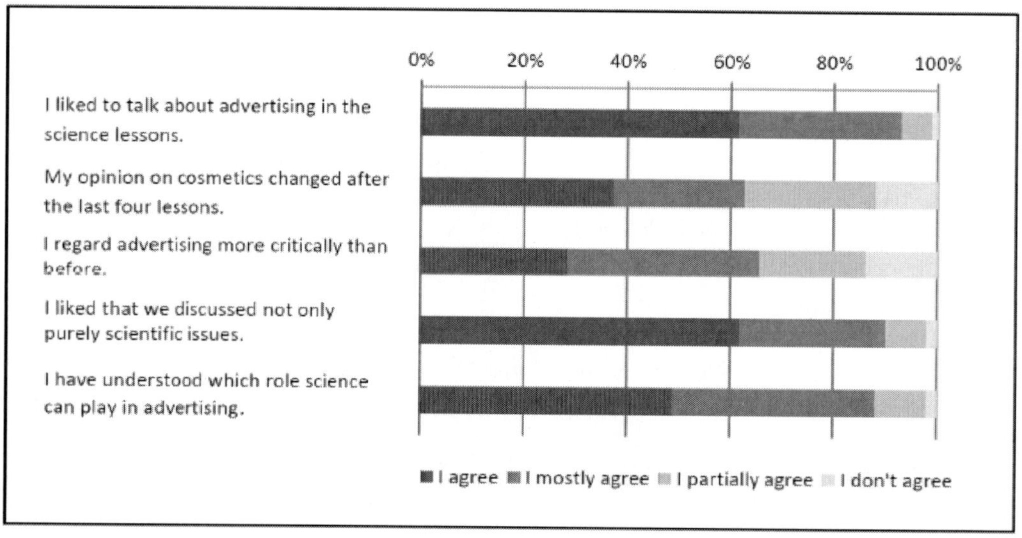

Figure 5. Findings from selected Likert items.

In classroom discussions they tend to fall back on analysis strategies learned in other subjects, including the evaluation of language phenomena, especially at the beginning of the teaching module.

Conclusion

Advertising influences students' consumption behavior and can simultaneously influence their perception of science (Davies, Spencer, Quinn and Gerhardstein, 2002; Dixon, Scully, Wakefield, White and Crawford, 2007).

McSharry and Jones (2002) have pointed out that 'advertisements could prove to be extremely useful in increasing the relevance of science education to children' (p. 496). However, such ideas remain largely neglected in the science education literature with regard to employing learning both through and about advertising.

Our study suggests that teaching using advertising as both a tool and as a topic in the science classroom of the future may potentially better science education. This includes bettering science's contribution to the development of civic scientific media literacy in the next generation of learners as a prerequisite for their role as critical consumers. By providing a theoretical background which encompasses research findings from different domains we have shown that the application of advertising is justified and can be conducted in multiple ways. The objectives encompass a focus on both scientific inquiry and content knowledge, while incorporating elements which contribute to critical media literacy, socio-scientific communication capabilities and decision-making skills.

Advertising is an authentic medium and in many cases can also prove to be motivating. Dealing with advertising opens up new opportunities for the social contextualization of learning science and technology content. Innovative and student-oriented pedagogies, some of which were presented in this chapter, can help to counteract the perceived lack of relevance which many students express with regard to the sciences.

Several teaching strategies have been identified, based on the few available suggestions in the literature for effectively addressing advertising in the science classroom. Advertising approaches can be categorized according to the learning outcomes. These can be further differentiated between by separating learning *with* advertising from learning *about* advertising. The second aspect needs to be promoted more strongly, since it is currently underrepresented in both science education research and practice.

The student feedback shows that most pupils have a positive attitude towards learning about advertising in the science classroom. They regard advertising as a legitimate and suitable topic for the science classroom. Nevertheless, the classroom observations in particular revealed that students are not used to media education activities in science education. Moreover, older students seem to hold to a certain image of science which often does not include (or even allow) societal aspects.

Students need to be shown that science does include such aspects. In our opinion there is widespread need for more thorough implementation of socio-scientific issues in the science classroom, which directly includes advertising.

Research on teacher viewpoints shows that the ground has already been prepared for at least some of the teachers in our sample to start integrating science learning and media education more thoroughly. Our findings suggest that the majority of science teachers in the sample already have ideas on how to integrate advertising in the science classroom. Most of them also have done so in practice. However, this implementation is neither regular nor systematic and aims more at contextualizing subject matter learning than on contributing to media literacy. This is in line with the findings of other studies on teaching practices using science-related news media (e.g. Jarman and McClune, 2002).

There is a need for corresponding teacher education programs which address both traditional and newer media types, while employing advertising as both a tool and a topic for teaching and learning. This will allow both educators and educational systems to keep pace with the younger generation and the realities of their everyday lives. Curriculum development encompassing new media types is also necessary.

In terms of further research, there is a great need for more evidence. This is particularly true in the area covering the effects of including advertising in science classes and the concrete effects on student learning outcomes. In conducting such further, more specified research, we anticipate that using advertising in science education will reveal itself as an extremely important teaching and learning approach, that new, innovative pedagogies will result, and that civic scientific media literacy in the modern age will be promoted and bettered in the long term.

References

Albe, V. (2008). When scientific knowledge, daily life experience, epistemological and social considerations intersect: students' argumentation in group discussions on a socio-scientific issue. *Research in Science Education*, 38(1), 67–90.

An, S., Jin, H. S. and Park, E. H. (2014). Children's advertising literacy for advergames: perception of the game as advertising. *Journal of Advertising*, 43(1), 63-72.

Belova, N. and Eilks, I. (2014). Using advertisings to introduce inquiry and societal oriented science education. *Centre for Educational Policy Studies Journal*, 4(1), 32-49.

Bever, T. G., Smith, M. L., Bengen, B., and Johnson, T. G. (1975). Young viewers' troubling response to TV ads. *Harvard Business Review*, 53(6), 109–120.

Biddle, B. J. and Anderson, D. S. (1986). Theory, methods, knowledge and research on teaching. In: M. C. Wittrock (Ed.), *Handbook of Research on Teaching* (3rd ed., pp. 230–252). New York, NY: Macmillan.

Brossard, D. and Shanahan, J. (2006). Do they know what they read? Building a scientific literacy measurement instrument based on science media coverage. *Science Communication*, 28(1), 47-63.

Brucks, M., Armstrong, G. M. and Goldberg, M. E. (1988). Children's use of cognitive defenses against television advertising: A cognitive response approach. *Journal of Consumer Research*, 14(4), 471-482.

Burmeister, M. and Eilks, I. (2012). An example of learning about plastics and their evaluation as a contribution to Education for Sustainable Development in secondary school chemistry teaching. *Chemistry Education Research and Practice*, 13(2), 93-102.

Burrows, W. R. (1997). The great paper towel race. *Science Activities*, 34(3), 32-37.

Butter, E. J., Popovich, P. M., Stackhouse, R. H., and Garner, R. K. (1981). Discrimination of television programs and commercials by preschool children. *Journal of Advertising Research*, 21(2), 53–56.

Byrd-Bredbenner, C. (2002). Saturday morning children's television advertising: a longitudinal content analysis. *Family and Consumer Sciences Research Journal*, 30(3), 382-403.

Chang Rundgren, S.-N. and Rundgren, C.-J. (2014). SSI pedagogic discourse: embracing scientific media literacy and ESD to face the multimedia world. In: I. Eilks, S. Markic and B. Ralle (Eds.), *Science education research and education for sustainable development* (pp. 157-168). Aachen, Germany: Shaker.

Christensen, C. K. (2011). Young adults' accounts of scientific knowledge when responding to a television news report of contested science. *International Journal of Science Education, Part B: Communication and Public Engagement*, 1(2), 115-145.

Connell, P. M., Brucks, M. and Nielsen, J. H. (2014). How childhood advertising exposure can create biased product evaluations that persist into adulthood. *Journal of Consumer Research*, 41(1), 119-134.

Constandinidou-Semoglou, O. (2007). Early childhood education and adult-oriented advertising discourse. *European Early Childhood Education Research Journal*, 15(3), 329-341.

Cox, T., McCollum, J. S. and Watkins, R. K. (1938). Science Claims in magazine advertising. *Science Education*, 22(1), 14-19.

Davies, P. G., Spencer, S. J., Quinn, D. M., and Gerhardstein, R. (2002). Consuming images: How television commercials that elicit stereotype threat can restrain women academically and professionally. *Personality and Social Psychology Bulletin*, 28(12), 1615-1628.

Dhingra, K. (2003). Thinking about television science: How students understand the nature of science from different program genres. *Journal of Research in Science Teaching*, 40(2), 234-256.

Dixon, H. G., Scully, M. L., Wakefield, M. A., White, V. M., and Crawford, D. A. (2007). The effects of television advertisements for junk food versus nutritious food on children's food attitudes and preferences. *Social Science and Medicine*, 65(7), 1311-1323.

Eilks, I., Nielsen, J. A. and Hofstein, A. (2014). Learning about the role of science in public debate as an essential component of scientific literacy. In: A. Tiberghien, C. Bruguière, P. Clément (Eds.), *Topics and trends in current science education* (pp. 85-100). Dordrecht: Springer.

Elliott, P. (2006). Reviewing newspaper articles as a technique for enhancing the scientific literacy of student teachers. *International Journal of Science Education*, 28(11), 1245-1265.

European Commission (2007). *Media literacy. Active citizenship in today's information society.* Retrieved from http://ec.europa.eu/culture/media/media-literacy/index_en.htm.

Feierabend, T. and Eilks, I. (2011). Teaching the societal dimension of chemistry using a socio-critical, problem-oriented lesson plan based on bioethanol usage. *Journal of Chemical Education*, 88(9), 1250-1256.

Friestad, P. M. and Wright, P. (1994). The persuasion knowledge model: How people cope with persuasion attempts. *Journal of Consumer Research*, 21(1), 1–31.

Halkia, K. and Mantzouridis, D. (2005). Students' views and attitudes towards the communication code used in press articles about science. *International Journal of Science Education*, 27(12), 1395–1411.

Hall, H. L. and Aimone, L. H. (2008). *Teacher's workbook and teacher's guide for high school journalism.* New York, NY: Rosen Publishing.

Haney, J. J., Czerniak, C. M. and Lumpe, A. T., (1996). Teacher beliefs and intentions regarding the implementation of science education reform strands. *Journal of Research in Science Teaching*, 33(9), 971–993.

Hang, H. and Auty, S. (2011). Children playing branded video games: The impact of interactivity on product placement effectiveness. *Journal of Consumer Psychology*, 21(1), 65–72.

Hansen, A. (1994). Journalistic practices and science reporting in the British press. *Public Understanding of Science*, 3(2), 111–134.

Hanuscin, D. (2002). Names and Claims: Is it science or spin? *Science Scope*, 25(6), 36-38.

Hartnett, R. (2006). *Small business, big opportunity*. Melbourne, Australia: Sensis.

Hattie, J. A. C. (2009). *Visible learning. A synthesis of over 800 meta-analyses relating to achievement.* London, UK: Routledge.

Hobbs, R. (2003). Understanding teachers' experiences with media literacy in the classroom. In: B. Duncan and K. Tyner (Eds.), *Visions/revisions: Moving forward with media education* (pp. 100–108). Madison, WI: National Telemedia Council.

Hobbs, R. and Jensen, A. (2009). The past, present, and future of media literacy education. *Journal of Media Literacy Education*, 1(1), 1-11.

Hofstein, A., Eilks, I. and Bybee, R. (2011). Societal issues and their importance for contemporary science education: a pedagogical justification and the state of the art in Israel, Germany and the US. *International Journal of Science and Mathematics Education*, 9(6), 1459-1483.

Holbrook, J. and Rannikmäe, M. (2007). The nature of science education for enhancing scientific literacy. *International Journal of Science Education*, 29(11), 1347-1362.

Holmes, H. W., et al. (Eds.) (1947). *Fundamental Education, Common Ground for All Peoples. Report of a Special Committee to the Preparatory Commission of the United Nations Educational, Scientific and Cultural Organization, Paris, 1946.* New York, NY: Macmillan.

Hopkins, L. T. (1925). A study of magazine and newspaper science articles with relation to courses in sciences for high schools. *School Science and Mathematics*, 25(8), 793–800.

Hove, T., Paek, H.-J. and Isaacson, T. (2011). Using adolescent eHealth literacy to weigh trust in commercial web sites: The more children know, the tougher they are to persuade. *Journal of Advertising Research*, 51(3), 524-537.

Jarman, R. and McClune, B. (2001). Use the news: A study of secondary teachers' use of newspapers in the science classroom. *Journal of Biological Education*, 35(2), 69–74.

Jarman, R. and McClune, B. (2002). A survey of the use of newspapers in science instruction by secondary teachers in Northern Ireland. *International Journal of Science Education*, 24(10), 997–1020.

Johnson, M. and Pigliucci, M. (2004). Is knowledge of science associated with higher skepticism of pseudoscientific claims? *The American Biology Teacher*, 66(8), 536-548.

Kachan, M. R., Guilbert, S. M. and Bisanz, G. L. (2006). Do teachers ask students to read news in secondary science? *Evidence from the Canadian context Science Education*, 90 (3), 496–521.

Klosterman, M. L., Sadler, T. D. and Brown, J. (2012). Science teachers' use of mass media to address socio-scientific and sustainability issues. *Research in Science Education*, 42 (1), 51-74.

Koelsche, C. (1965). Scientific literacy as related to the media of mass communication. *School Science and Mathematics*, 65(8), 719–725.

Kolstø, S. D. (2001). 'To trust or not to trust ...': Pupils' ways of judging information encountered in a socio-scientific issue. *International Journal of Science Education*, 23(9), 877–901.

Landesinstitut für Schule Bremen (2012). *Bildungsplan Medienbildung*. Retrieved from http://www.lis.bremen.de/sixcms/media.php/13/2012_bpmedien_aktuell.36056.pdf.

Lauri, M. A., Borg, J., Günnel, T., and Gillum, R. (2010). Attitudes of a sample of English, Maltese and German teachers towards media education. *European Journal of Teacher Education*, 33(1), 79-98.

Lemke, J. (2004). The literacies of science. In: E. W. Saul (Ed.), *Crossing borders in literacy and science instruction: Perspectives on theory and practice* (pp. 33–47). Newark/ Arlington, NJ: International Reading Association/National Science Teachers Association.

Livingstone, S. and Helsper, E. J. (2006). Does advertising literacy mediate the effects of advertising on children? A critical examination of two linked research literatures in relation to obesity and food choice. *Journal of Communication*, 56(3), 560–584.

Marks, R. and Eilks, I. (2009). Promoting scientific literacy using a socio-critical and problem-oriented approach to chemistry teaching: concept, examples, experiences. *International Journal of Environmental and Science Education*, 4(3), 231-245.

Marks, R., Otten, J. and Eilks, I. (2010). Writing news spots about chemistry – A way to promote students' competencies in communication and evaluation. *School Science Review*, 92(339), 99-108.

Marks, R., Stuckey, M., Belova, N., and Eilks, I. (2014). The societal dimension in German science education – From the tradition to selected cases and recent developments. *Eurasia Journal of Mathematics, Science and Technology Education*, 10(4), 285-296.

Martin, W. E. (1945). A chronological survey of published research studies relating to biological materials in newspapers and magazines. *School Science and Mathematics*, 45 (6), 543–550.

Mayring, P. (2000). *Qualitative content analysis. Forum Qualitative Social Research,* 1(2). Retrieved from http://www.qualitative-research.net/index.php/fqs/article/view/1089/2386.

McClune, B. and Jarman, R. (2012). Encouraging and equipping students to engage critically with science in the news: what can we learn from the literature? *Studies in Science Education*, 48(1), 1-49.

McSharry, G. and Jones, S. (2002). Television programming and advertisements: Help or hindrance to effective science education? *International Journal of Science Education*, 24 (5), 487–497.

Nisbet, M. C., Scheufele, D. A., Shanahan, J., Moy, P., Brossard, D., and Lewenstein, B. V. (2002). Knowledge, reservations or promise? A media effects model for public perceptions of science and technology. *Communication Research*, 29(5), 584-608.

Owen, L., Lewis, C., Auty, S., and Buijzen, M. (2013). Is children's understanding of nontraditional advertising comparable to their understanding of television advertising? *Journal of Public Policy and Marketing*, 32(2), 195-206.

Palmer, E. L. and McDowell, C. N. (1979). Program/commercial separators in children's television programming. *Journal of Communication*, 29(3), 197–201.

Pitrelli, N., Manzoli, F. and Montolli, B. (2006). Science in advertising: Uses and consumptions in the Italian press. *Public Understanding of Science*, 15(2), 207-220.

Preston, C. (2006). Children's interaction with adult orientated advertising. *The Marketing Review*, 6(1), 73-84.

Priest, S. (1999). Structuring public debate on biotechnology. Media frames and public response. In: E. Scanlon, E. Whitelegg and S. Yates (Eds.), *Communicating science. Contexts and channels*. (pp. 240-251). London, UK: Routledge.

Robertson, T. S. and Rossiter, J. R. (1974). Children and commercial persuasion: an attribution theory analysis. *Journal of Consumer Research*, 1(1), 13–20.

Rozendaal, E., Buijzen, M. and Valkenburg, P. M. (2009). Do children's cognitive defenses reduce their desire for advertised products? *Communications: The European Journal of Communication Research*, 34, 287–303.

Rozendaal, E., Buijzen, M. and Valkenburg, P. M. (2012). Think-aloud process superior to thought-listing in increasing children's critical processing of advertising. *Human Communication Research*, 38(2), 199-221.

Rundgren, C.-J., Chang-Rundgren, S.-N., Tseng, Y.-H., Lin, P.-L., and Chang, C.-Y. (2012). Are you SLiM? –Developing an instrument for civic scientific literacy measurement (SLiM) based on media coverage. *Public Understanding of Science*, 21(6), 759-773.

SACE Board of South Australia (2014): Stage 2 media studies. *Subject operational information for 2014.* Retrieved from http://www.sace.sa.edu.au/documents/652891/704861/Stage+2+Media+Studies+-+Subject+Operational+Information+for+2014.pdf/49bf5dd2-cac5-47a5-912e-e793a3b6d760?version=1.4.

Sadler, T. D. (2004). Informal reasoning regarding socioscientific issues: A critical review of research. *Journal of Research in Science Teaching*, 41(5), 513–536.

Sadler, T. D. (2011). *Socio-scientific issues in the classroom.* Heidelberg: Springer.

Scheibe, C. and Rogow, F. (2012). *The teacher's guide to media literacy.* Thousand Oaks, CA: Corwin.

Sjöström, J. (2013). Towards Bildung-oriented science education. *Science and Education*, 22 (7), 1873-1890.

Stuckey, M., Hofstein, A., Mamlok-Naaman, R., and Eilks, I. (2013). The meaning of 'relevance' in science education and its implications for the science curriculum. *Studies in Science Education*, 49(1), 1-34.

Stuckey, M., Lippel, M. and Eilks, I. (2012). Sweet chemistry: Learning about natural and artificial sweetening substances and advertising in chemistry lessons. *Chemistry in Action*, 98, 36-43.

UNESCO (1982). *Challenge of media education (The Grunwald Document).* Retrieved from http://www.medialit.org/reading-room/challenge-media-education-grunwald-document.

UNESCO (2006). *Media education. A kit for teachers, students, parents and professionals.* Retrieved from http://unesdoc.unesco.org/images/0014/001492/149278e.pdf.

UNESCO (2011). *Media and information literacy. Curriculum for teachers.* Retrieved from http://unesdoc.unesco.org/images/0019/001929/192971e.pdf.

Van Driel, J., Bulte, A. M. W. and Verloop, N. (2007). The relationships between teachers' general beliefs about teaching and learning and their domain specific curricular beliefs. *Learning and Instruction*, 17(2), 156-171.

Villani, S. (2001). Impact of media on children and adolescents: A 10-Year review of the research. *Journal of the American Academy of Child and Adolescent Psychiatry*, 40(4), 392-401.

Ward, S. (1972). Children's reactions to commercials. *Journal of Advertising Research*, 12 (2), 37–45.

In: New Developments in Science Education Research
Editor: Nathan L. Yates
ISBN: 978-1-63463-914-9

Chapter 3

COLLABORATIVE CURRICULUM DEVELOPMENT OF A TEACHING AND LEARNING MODULE ON BIONICS BASED ON INNOVATIVE ICT TECHNOLOGY

Moritz Krause, Dörte Ostersehlt and Ingo Eilks
Institute of Science Education, University of Bremen, Germany

ABSTRACT

Bionics is a topic in which learning about modern applications of science with reflection on current changes in science and technology can be combined. However, the availability of innovative teaching and learning materials on bionics is still rather limited. This chapter addresses a curriculum development, classroom-based research project which follows the model of Participatory Action Research in science education. In a two-year timespan, two university science educators and four high school teachers collaboratively developed, tested and revised a lesson plan for teaching bionics in lower secondary science education. The central pedagogical innovation was to incorporate all teaching and learning activities in a multimedia-based learning environment. This environment combines different sorts of media and activities such as explanatory texts, learning videos, physical models, inquiry-type experiments, and assessment tasks. The classroom scenario with the corresponding learning environment is described. A reflection upon the experience gained during the collaborative process of curriculum development is also provided. Selected findings are discussed.

1. INTRODUCTION

In the 1990s, the establishment of the Internet created a new platform for information storage. It also provided a new way to exchange information, which at that time was mainly used by the military and research institutes (Tuvi & Nachmias, 2003). Over the last two decades the situation has changed dramatically (Songer, 2007). Today, the Internet connects the entire world and makes information available in a manner never before seen. For many people the Internet became their first information resource (Murray-Rust, Rzepa, Tyrrell & Zhang, 2004). The Internet has achieved a central role in all domains of modern life. It also has become part of today's students' lives and an important source of information for learning, too (Frailich, Kesner & Hofstein, 2009).

As ICT took on increasing importance in all domains of modern societies, it also changed schools and the lives of teachers (Songer, 2007). At least in industrialized countries, most modern teachers use the Internet and related media such as learning tools, animations, or databases to improve teaching and learning (Dori, Rodrigues & Schanze, 2013). Improvements in hard- and software make the use of ICT for teaching easier and more broadly available every year.

In the last thirty years many different strategies and pedagogies have been suggested for employing computer-based learning environments and the Internet in education in general and in science education in particular (e.g. Linn, 1998; Rodrigues, 2010; Songer, 2007). New developments in hardware like smartphones, tablet-PCs and interactive whiteboards, as well as continuous improvement in the available software, cause constant challenges in innovating educational practices through modern ICT. The same holds true for research in this area (Dori et al., 2013).

Research has revealed that the use of ICT can have benefits for learning, for example using animations and visualizations (Williamson & Abraham, 1995; Falvo, 2008). Yet any benefit of using ICT in education is not self-evident. ICT is only supportive of learning if many basic principles in its use are taken into consideration. Mayer (2003) discovered that digital media in education is only supportive of learning if: 1) texts and figures are structured in a well-attuned combination, 2) if no irrelevant information distracts the attention of the learner, 3) texts and the corresponding visual information are situated close to one another, and 4) the quality of textual and visual information satisfies the communication experiences of the learner. However, the didactic structure of information, the proper visualization of content and its textual embedding are also vital for multimedia-based learning success (Barak & Dori, 2011; Kerres, 2000). As in any other media and learning environment, it is important that the information is well structured and reliable in terms of content and potential misconceptions, yet nevertheless provokes activity and cognitive engagement in the learner (Eilks, Witteck & Pietzner, 2010).

This paper discusses a curriculum development project taking all these thoughts into account. It presents an approach for using a new software (PREZI) in connection with the broad availability of hand-held and laptop computers to promote inquiry-based science education. PREZI allows for interactive ways of presenting information to the learner. Information can be structured in a layered and networked fashion. It can also be organized so that the learner can choose a personal learning path and select the pace of dealing with the material. Nevertheless, the teacher can also suggest trajectories for learning, which the students are free to accept or reject. Content can be illustrated using a whole set of different media types. As a sym-medium PREZI allows the combination of texts, videos, animations, supportive materials on experiments, Internet search results and practical inquiries. This paper describes one example case of curriculum development based on the use of PREZI technology to support inquiry-based science education. The development process was guided by the model of Participatory Action Research in science education as suggested by Eilks and

Ralle (2002). The case study will be illustrated through an example taken from junior high school science on the topic of bionics.

2. Curriculum Development by Participatory Action Research in Science Education

The curriculum development reported in this chapter was part of the EU-funded project PROFILES (Professional Reflection-Oriented Focus on Inquiry Learning and Education through Science). PROFILES aims to foster inquiry-based learning and a stronger societal orientation of science education via investment in the continuous professional development (CPD) of teachers (Bolte et al., 2011). In the PROFILES group at the University of Bremen, Germany, a CPD approach of collaborative curriculum development was selected to develop and implement new teaching and learning environments following the PROFILES philosophy. The developmental process was carried out using Participatory Action Research (Eilks and Ralle, 2002). In the past this model has proven itself to be fruitful for both research-based curriculum development (Marks & Eilks, 2010; Eilks & Feierabend, 2013) and for contributing to teacher CPD (Eilks, 2014; Mamlok-Naaman & Eilks, 2012).

Participatory Action Research in science education is a cyclical process of curriculum innovation and classroom-based research in a partnership of practicing teachers and science educators (Eilks & Ralle, 2002). The foci and initiatives towards practice improvement are negotiated jointly by all the persons involved. For collaboration the group meets about once a month for discussion, work on classroom materials and exchange of experiences or accompanying research findings (Eilks, Mamlok-Naaman & Rauch, 2012).

The learning environment on bionics was created over a period of two consecutive school years. Four science teachers and two university science educators were involved. At times, the group was supported by Master's degree candidates, who had aided the research and development process as part of their Masters of Education thesis work. The group’s intention was to carry out curriculum development on bionics at the junior high school level. The primary focus was defined by developing classroom innovations around the lotus effect. Initial ideas were collected and selected. In the next three sessions a learning path was structured and potential classroom experiments were also tested. Piece by piece, a first lesson plan was developed. After completing the first version of this part of the course, the lesson plan on the lotus effect was tested in the classroom. Data from teachers and students were collected with different oral and written tools. The data served to refine all of the teaching and learning activities and the corresponding materials. It also helped to uncover the participants' perceptions of the developed lesson plan and the impact on the students (Krause et al., 2014).

In a second cycle of development, the possibility of embedding all the classroom materials and activities in a multimedia learning environment to integrate new media and experiments in the classroom was discussed. A website was established and different multimedia tools were created for presentation, learner support, and assessment. To enrich the learning environment learning videos were created so that students had an opportunity to work on the materials not just in the classroom, but also at home. In addition, a second lesson plan on the "Gecko on the wall" was structured. These materials were also initially developed as traditional print materials, but later implemented into the multimedia-based learning environment. Finally, a third lesson plan was developed covering the topic "From the bird to the aircraft". All the materials were then further tested, evaluated and improved in various classes. Feedback from students and teachers led to a final revision of the lesson plan and the corresponding multimedia learning environment.

3. PREZI – A New Tool for the Creation of a Learning Environment on Bionics

3.1. PREZI

PREZI is presentation software which was developed in 2007 and presented to the public in 2009 by the Hungarian artist, Adam Somlai-Fischer, and the computer scientist, Peter Halacsy. PREZI can be seen as a more elaborated, interactive version of traditional presentation software programs like Microsoft Power Point. In contrast to Power Point, PREZI allows users to present information in a tighter, more networked fashion, which includes also a more dynamic appearance.

The idea behind PREZI is based on a desktop (see Figure 1). Different slides containing information can be set up in this area and arranged quite freely. They can also be designated to follow a specific order of appearance. The slides on the desktop can also be assigned to different layers. This means that slides can not only be arranged in specific sequences, but also can be stacked by the author in the case that information needs to be deepened and expanded. Thanks to this feature PREZI allows three-dimensional structuring of information. In practice, PREZI technology allows users to follow information in a specific sequence, but then stop and zoom in on any information where the organizer has provided extra materials.

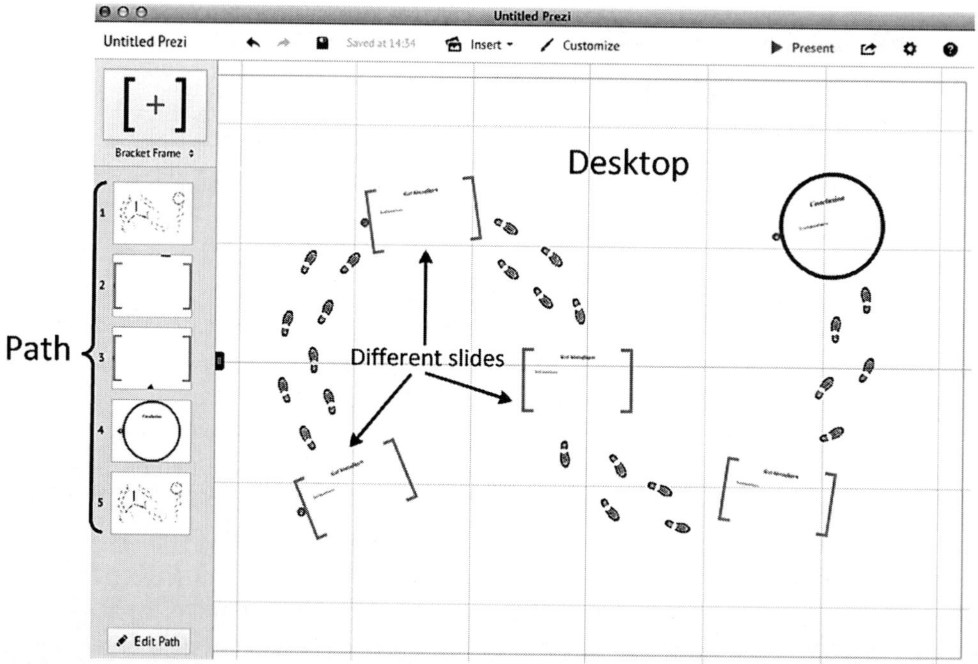

Figure 1. Sketch of a PREZI user interface.

PREZI allows authors to present slides in selected sequences just like Power Point. However, it is also possible to jump from one slide on the desktop to another. The user can therefore decide to follow a suggested learning path, leave it, or ignore parts of it. Each slide allows the integration of different kinds of media such as texts, pictures, videos or animations, which can be directly accessed by the PREZI environment.

A highly networked and multimedia supported information resource can be created when using PREZI, either with the accompanying software or as a stand-alone tool. The software also generates a dynamic record of its use, because the different slides are presented as if the user was walking through them. This impression is similar to the way most students today feel when using smartphones or tablet PCs.

3.2. PREZI to Create Learning Environments

PREZI was originally created as a presentation tool. It was not meant to be used for student-centered learning environments. However, this project indicates that such a networked, multimedia-based tool for structuring information is highly suitable for creating computer-based learning environments. PREZI seems to be well-fitted to presenting content matter and various media forms for self-directed, autonomous or cooperative learning alone or in groups. It also seems to have potential to select criteria for the effective use of ICT in learning. In other words it can structure activity by "*offering the students responsibility, choice and opportunities for active learning*" and by "*prompting students to think about new concepts and relationships, to participate in discussions, to analyze critically data and information, and to focus on research tasks*" as discussed in Dori et al. (2013) based on Osborne and Hennessy (2003).

PREZI's ability allows learners to follow a prescribed structure or to freely move between the parts opens to door to the pupil eventually arriving at the stage of discovery learning. This freedom to interact with content is much higher than in conventional presentations. PREZI environments can also be distributed by the Internet and be used with every conventional web browser. This means that learning becomes possible beyond the classroom and can also help with homework assignments or in individual preparation for assessments and tests.

Although PREZI technology is quite developed today, it also has a few limitations, since a few tools for creating learning environments are still missing. This is especially true when it comes to tools for assessment and self-assessment, because multiple choice items, cloze tests, etc. are not available. However, embedding PREZI in a broader learning environment allows the teacher to combine different learning phases both with PREZI and with assessment exercises occurring between the phases. In this way, different components (PREZI, cloze tests, etc.) can be combined (Figure 2).

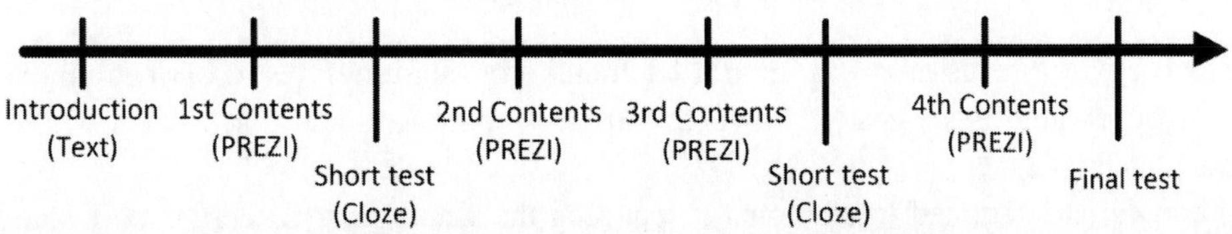

Figure 2. Potential structure of PREZI-based learning environments on bionics.

The students can learn, discuss, and assess their learning autonomously and in a self-directed manner, which has proven itself to be both motivating and effective in computer-supported learning environments (Rodrigues, 2010).

3.3. Using PREZI for Learning about Bionics

The teaching and learning environment on bionics offers an interactive learning path with chances for inquiry-based learning. It introduces content and concepts, as well as leading to learning through experiments and models. The learning environment consists of three sub-modules: "The lotus effect", "The gecko on the wall" and "From the bird to the aircraft". Figure 3 shows the first page of the learning module "The lotus effect". Each part of the lesson plan consists of a PREZI learning offering with various media elements such as videos, experimental activities, and modeling activities embedded in it. The assessments between PREZI phases provide the learners with feedback on their achievements. A final quiz can also be used for summative assessment.

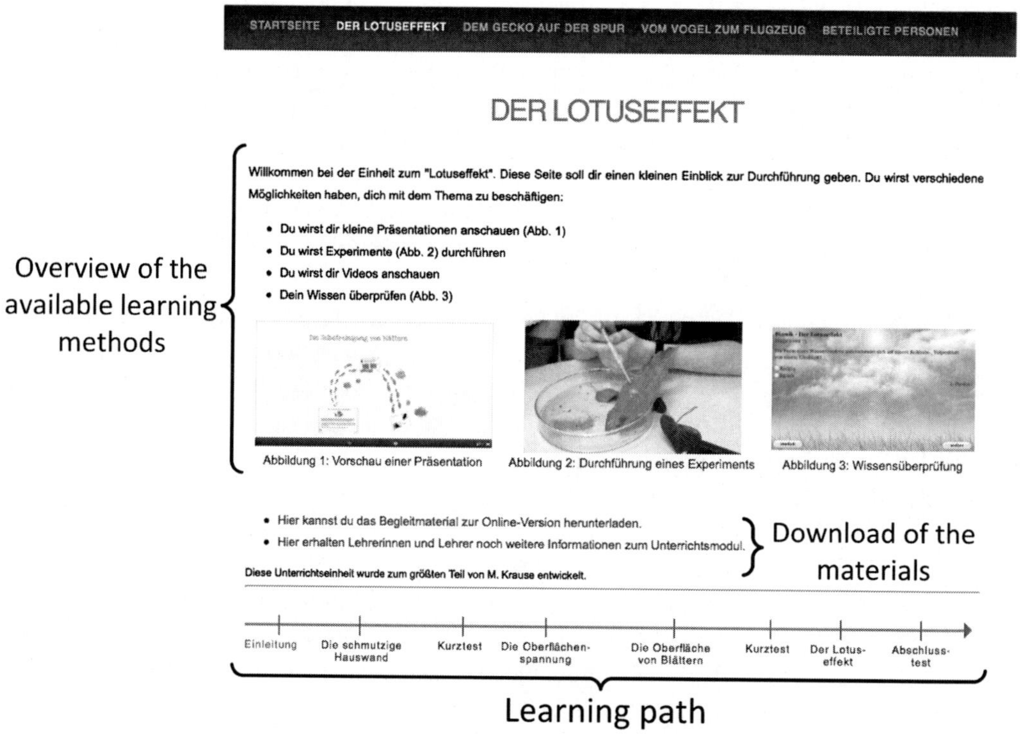

Figure 3. Overview of the learning environment on the lotus effect.

PREZI environments are the central elements for learning. As described above, a trajectory for learning is suggested, however individual pathways and discovery learning are also possible (Figure 4).

All content is integrated into a context spanning the whole unit. At certain way stations, problems must be solved through experiments and model creation. Hints and ideas are provided for the practical work and for self-directed creation of scientific models. In the sense of guided inquiry learning (Tafoya, Sunal & Knecht, 1980), no prescribed manuals are offered for the practical working phases. The students only receive the problem, the necessary materials, and structured aids in order to experience cooperative problem solving in small groups (Figure 5).

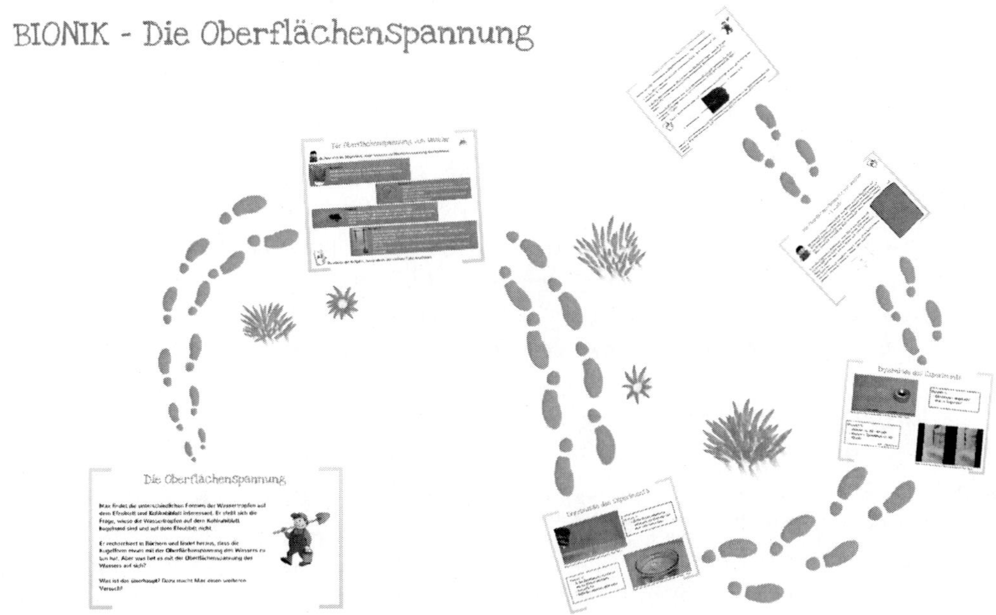

Figure 4. Overview of the learning pathway on surface tension.

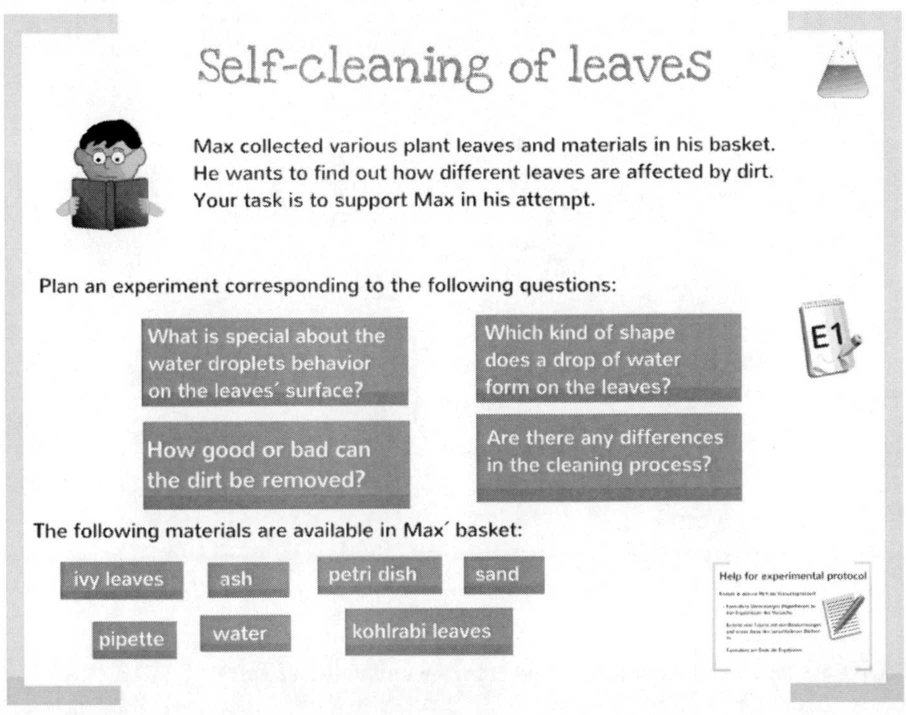

Figure 5. Translation of the instructions for guided inquiry experimentation.

3.4. The Use of Videos in the PREZI Learning Environment

Students are presented with many suggestions for experiments and modeling activities in the PREZI environments. However, experiments and explanations are also given in videos on the PREZI desktop. The pupils can use these videos by different strategies. Most students choose to first perform the experiments then check the videos to confirm their observations. More insecure learners first watch the video to see how to perform the experiment before carrying it out on their own. Pupils can not only decide what sequence of experiments and videos to follow, they can also interact with the videos by stopping, fast-forwarding or rewinding them according to their cognitive needs (Chandler, 2004; Mayer & Chandler, 2001).

There is also a second layer of videos embedded in the program. Four learning videos were created for this purpose. Learning videos are different from experimentation videos, because they encompass not only experiments or observations, but organize them in an explanatory context. Research shows that the ability to stop and watch individual scenes again helps building both declarative as well as procedural knowledge in learners (Merkt et al., 2011; Schwan & Riempp, 2004). One learning video focuses on the question "Why does the gecko adhere to the wall?". It explains the phenomenon step-by-step so that even students with no prior knowledge can be introduced to Van der Waals forces as the main reason that the gecko sticks to the wall without falling. In "From the bird to the aircraft" the video (Figure 6) is part of a learning-at-stations scenario (Eilks, 2002). Students learn hands-on physical aspects at the various experimental stations in the classroom, before viewing a video summary to supplement their conclusions and/or correct them.

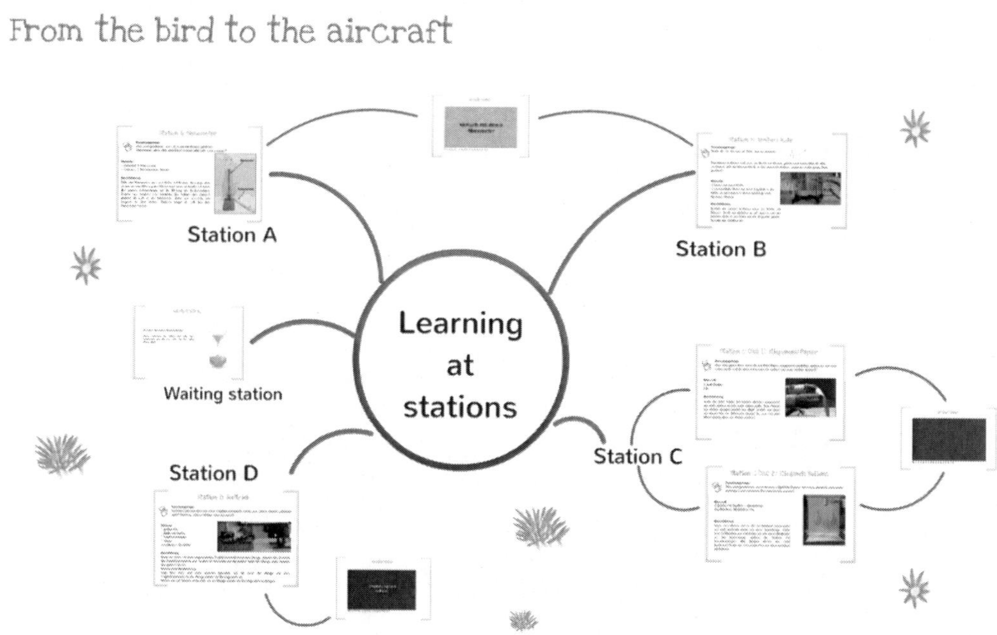

Figure 6. Displays the PREZI work station for “From the bird to the aircraft”.

3.5. Using the Learning Environment in Class

The interactive learning environment is designed for multiple uses in class. Combinations of in-class learning with homework assignments are possible. Individual work, cooperative learning and common phases of exchange by the whole class can easily be incorporated. Whenever applicable, the students leave the computers and perform experiments at laboratory stations when they need to (Figure 7). Print materials are used to document any hands-on activity or content derived from the computer environment. The learning experience becomes a mix of new content, experiments and exercises. Since the learning environment is accessible on-line, pupils also use it to prepare themselves for class work or assessments. The learning platform is also open to the general public to be used in informal or non-formal educational settings.

Figure 7. Students use computers in the front of the picture and experimentation tables in the rear of the classroom.

4. Experiences from the Classroom

The lesson plan has so far been tested in several 7th and 8th grade classes at various schools (n = 149; age range 14-15). The lessons were evaluated through teacher feedback and student questionnaires. During the development and testing cycles, different research questions were taken into focus. The accompanying research encompassed: 1) the use and application of the multimedia learning environment in the classroom, 2) participants' perceptions of the learning videos, 3) pupils' learning progress on the topic of bionics, and 4) the development of inquiry-based skills in the students.

Data from the questionnaires shows that the students generally have great interest in bionics even before the lesson plan begins. This interest is generally maintained during the unit and students thoroughly learn the essential aspects of bionics. Declarative knowledge levels significantly improved when comparing the pre- and post-test results. Procedural knowledge about scientific inquiry also significantly improved. These findings support research that multimedia learning environments have the potential to promote knowledge and competency if connected to inquiry-based learning guided by activities and media (Handelsman et al. 2004; Dori, Hult, Breslow & Belcher, 2007). The learning videos showed that they could help students understand complex knowledge as has already been suggested by Merkt et al. (2011) and Schwan and Riempp (2004). The gecko videos aided students' understanding of such concepts as adhesion and Van der Waals forces. The students stated that the integration of the videos helped them to better understand the content.

In the open questionnaire, students stated that the biggest difference compared to conventional science lessons was the high level of autonomy in learning. This coincided with the teachers' views. Pupils also mentioned the large number of experiments that they were able to conduct based on the computer learning environment. One student said: "We didn't obtain our tasks from the teacher, instead we had to work on the whole topic and the experiments on our own". Particularly the self-directed work with the computer and the experiments were seen by the students as the biggest benefits stemming from this form of lesson organization. The participants appreciated the opportunity to work at their own pace and to deal with the learning content individually or in small groups.

Likert questionnaires confirmed the positive perceptions of the open and self-directed learning atmosphere expressed in the answers to the open questions. All of the learners largely agreed that the learning environment was well-structured and that there were no problems working within it. Feedback showed that students viewed the learning environment as very supportive and feasible. More than 90% agreed that PREZI is user-friendly. About 80% of the students stated that the dynamic character of PREZI presentations makes the learning environment modern in appearance. The combination of the computer learning environment, experiments, and paper handouts was judged to be good by 90% of the students. Only a few of the students supported a more teacher-centered pedagogy or preferred having all of the materials in written format. About 95% agreed that they learned a lot with the aid of the new learning environment.

Overall, the feedback from students and teachers showed very positive perceptions of the lesson plan and the corresponding media items. Students rated as very positive both the specific computer work in connection with experiments and the use of a learning environment created with PREZI which integrated text, images and videos.

CONCLUSION

PREZI is a new and valuable tool for creating multimedia-based learning environments for inquiry-based science education. PREZI offers educators chances to create environments which allow for both structured and guided inquiry learning, as well as even more open forms of learning. The usability and the dynamic, modern appearance allow for self-directed and autonomous learning and is supported by the integration of a broad range of media and activities into one tool. This was highly appreciated by both students and teachers alike. This case study supports similar findings as previously reported in Krause, Kienast, Witteck and Eilks (2013).

As stated in the introduction, all of the benefits of using computers and ICT-based learning environments for learning are not self-evident. A suitable, feasible structure of content and activities are also necessary when it comes to using modern software tools for the benefit of the learners. PREZI can offer just such a platform for learning environments as has been shown here in the case of bionics. However, it was mainly the close cooperation of teachers and science educators, as well as the cyclical process of development, testing, reflection and revision, that secured a suitable curricular and methodological structure. This again supports the view that research-based knowledge and practical experience in the classroom compose the two ends of a spectrum of knowledge which is valuable for both curriculum development and classroom innovation (McIntyre, 2002).

The Participatory Action Research model for science education (Eilks & Ralle, 2002) showed again that this model has potential for developing practice- and user-friendly, feasible learning scenarios and materials (Marks & Eilks, 2010). It also confirmed, as suggested by Osborne and Hennessy (2003), that innovations are best built upon teachers' preexisting knowledge. This way helps to close the two-communities gap which tends to exist between in-service practitioners and university science education researchers (Mamlok-Naaman & Eilks, 2012). Although it was not specifically part of the initial focus of this case study, all of the planners learned a great deal about PREZI technology, how to use it, and how to exploit it for educational purposes.

A German version of the teaching and learning materials is available at: www.chemiedidaktik.uni-bremen.de/multimedia/bionik

We gratefully acknowledge the funding of the PROFILES project by the European Union under grant agreement number FP7-266589.

REFERENCES

Barak, M., & Dori, Y. J. (2011). Science education in primary schools: Is an animation worth a thousand pictures? *Journal of Science Education and Technology*, *20*, 608–620.

Bolte, C., Streller, S., Holbrook, J., Rannikmae, M., Mamlok-Naaman, R., Hofstein, A., Rauch, F. (2011). PROFILES: Professional Reflection-Oriented Focus on Inquiry based Learning and Education through Science. Proceedings of the European Science Educational Research Association (ESERA), Lyon, France, September 2011.

Chandler, P. (2004). The crucial role of cognitive processes in the design of dynamic visualizations. *Learning and Instruction*, *14*, 353-357.

Dori, Y. J., Hult, E., Breslow, L., & Belcher, J. W. (2007). How much have they retained? Making unseen concepts seen in a freshman electromagnetism course at MIT. *Journal of Science Education and Technology*, *16*, 299-323.

Dori, Y., Rodrigues, S., & Schanze, S. (2013). How to promote chemistry learning through the use of ICT. In I. Eilks, & A. Hofstein (Eds.), *Teaching chemistry – A studybook* (pp. 213-240). Rotterdam: Sense.

Eilks, I. (2002). "Learning at Stations" in secondary level chemistry lessons. *Science Education International, 13* (1), 11-18.

Eilks, I., (2014). Action Research in science education – From a general justification to a specific model in practice. In T. Stern, F. Rauch, A. Schuster, & A. Townsend (Eds.), *Bringing a different world into existence: action research as a trigger for innovations* (pp. 156-176). London: Routledge.

Eilks, I., & Feierabend, T. (2013). Developing the curriculum by Participatory Action Research – An interdisciplinary project on climate change. In T. Plomp & N. Nieveen (Eds.), *Educational design research: Introduction and illustrative cases* (pp. 321-338). Enschede: SLO.

Eilks, I., Mamlok-Naaman, R., & Rauch, F. (2012). Action Research for innovations and Continuous Professional Development (CPD) in science education. In: C. Bolte, J. Holbrook, F. Rauch (Eds.), *Inquiry-based science education in Europe: Reflections from the PROFILES project* (pp. 207-212). Berlin: FU Berlin.

Eilks, I., & Ralle, B. (2002). Participatory Action Research in chemical education. In B. Ralle, I. Eilks (eds.), *Research in Chemical Education - What does this mean?* (pp. 87-98). Aachen: Shaker.

Eilks, I., Witteck, T., & Pietzner, V. (2010). Using multimedia learning aids from the Internet for teaching chemistry – Not as easy as it seems?. In S. Rodrigues (ed.), *Multiple literacy and science education: ICTS in formal and informal learning environments* (pp. 49-69). Hershey: IGI Global.

Falvo, D. (2008). Animations and simulations for teaching and learning molecular chemistry. *International Journal of Technology in Teaching and Learning, 4,* 68-77.

Frailich, M., Kesner, M., & Hofstein, A. (2009). Enhancing students' understanding of the concept of chemical bonding by using activities provided on an interactive website. *Journal of Research in Science Teaching, 46*, 289-310.

Handelsman J., Ebert-May, D., Beichner, R., Bruns, P., Chang, A., Dehaan, R., Gentile, J., Lauffer, S., Stewart, J., Tilghman, S.M., Wodd, W.B. (2004). Scientific teaching. *Science*, 304, 521-522.

Kerres, M. (2000). Mediendidaktische Analyse digitaler Medien im Unterricht. *Computer und Unterricht, 10* (1), 26-28.

Krause, M., Kienast, S., Witteck, T., & Eilks, I. (2013). Tour de Chemie – Self-directed cooperative learning supported by computer assessment and feedback. *Chemistry Education Research and Practice, 14*, 345-353.

Krause, M., Ostersehlt, D., Mehrwald, T. ,Runden, H.-J., Mroske, S., & Eilks, I. (2014) Using PREZI-technology to promote inquiry-based learning on "bionics". In C. Bolte, J. Holbrook, R. Mamlok-Naaman & F. Rauch (Eds.), *Science teachers' continuous professional development in Europe, Cases from the PROFILES Project* (pp. 149-158). Berlin: FU Berlin.

Linn, M. (1998). The impact of technology on science instruction: Historical trends and current opportunities. In B. J. Fraser & K. G. Tobin (Eds.), *International handbook of science education* (pp. 265-295). Dordrecht: Kluwer.

Mamlok-Naaman, R., & Eilks, I. (2012). Action research to promote chemistry teachers' professional development – Cases and experiences from Israel and Germany. *International Journal of Mathematics and Science Education, 10*, 581-610.

Marks, R., & Eilks, I. (2010). Research-based development of a lesson plan on shower gels and musk fragrances following a socio-critical and problem-oriented approach to chemistry teaching. *Chemistry Education Research and Practice, 11*, 129-141.

Mayer, R. E. (2003). The promise of multimedia learning: Using the same instructional methods across different media. *Learning and Instruction, 13*, 125-140.

Mayer, R., & Chandler, P. (2001). When learning is just a click away: does simple user interaction foster deeper understanding of multimedia messages? *Journal of Educational Psychology, 93*, 390-397.

McIntyre, D. (2005). Bridging the gap between research and practice. *Cambridge Journal of Education*, 35, 357-382.

Merkt, M., Weigand, S., Heier, A., & Schwan, S. (2011). Learning with videos vs. learning with print: The role of interactive features. *Learning & Instruction, 21*, 687-704.

Murray-Rust, P., Rzepa, H. S., Tyrrell, S. M., & Zhang, Y. (2004). Representation and use of chemistry in the global electronic age. *Organic & Biomolecular Chemistry*, 2, 3192-3203.

Osborne, J., & Hennessy, S. (2003). Literature review in science education and the role of ICT: promise, problems and future directions. telearn.archives-ouvertes.fr/docs/00/19/04/41/PDF/osborne-j-2003-r6.pdf.

Rodrigues, S. (Ed.) (2010). *Multiple literacy and science education: ICTs in formal and informal learning environments*. Hershey: IGI Global.

Schwan, S., & Riempp, R. (2004). The cognitive benefits of interactive videos: learning to tie nautical knots. *Learning and Instruction, 14*, 293-305.

Songer, N. B. (2007). Digital resources versus cognitive tools: A discussion of learning science with technology. In S. K. Abell & N. G. Lederman (Eds.), *Handbook of research on science education* (pp. 471-492). Mahwah: Lawrence Erlbaum.

Tafoya, E., Sunal, D., & Knecht, P. (1980). Assessing enquiry potential: A tool for curriculum decision makers. *School Science and Mathematics, 80*, 43–48.

Tuvi, I., & Nachmias, R. (2003). A study of web-based learning environments focusing on atomic structure. *Journal of Computers in Mathematics and Science Teaching, 22*, 225-240.

Williamson, V. M., & Abraham, M. R. (1995). The effects of computer animation on the particulate mental models of college chemistry students. *Journal of Research in Science Teaching, 32*, 521-534.

In: New Developments in Science Education Research
Editor: Nathan L. Yates
ISBN: 978-1-63463-914-9

Chapter 4

Ask a Question! The Role of Student Questioning in University Science Classes

Patrícia Albergaria-Almeida*, Betina da Silva Lopes and Mariana Martinho
University of Aveiro, Research Centre Didactics and Technology in Education of Trainers, Portugal

Abstract

Students' questions play a crucial role in the learning process since questioning lies at the heart of scientific inquiry and meaningful learning. In the last decades several authors have emphasized the value of students' questions in science learning. A series of studies place the responsibility of questioning onto students rather than their teachers, and indicate that this benefits the quality of science teaching and learning, including at university level. In this chapter we propose to reflect on students' questions in university science classes, considering its relevance for learning and exploring questioning patterns associated to student learning styles and student gender, two topics that remain underexplored.

1. What does it Take to Ask Questions and Why Should They Be (More) Popular at (University) Science Classes?

The usefulness of questions for the teaching and learning processes is recognised by many (Almeida, 2007; Chin and Osborne, 2008; Dillon, 1988; Graesser and Olde, 2003; Lopes, 2013; Moreira, 2012; Van der Meij, 1994). However when asked to justify the importance of questions for teaching and learning the answer is not always as easy as it may seem at first sight (Olney, Graesser, and Person, 2012).

The first step towards understanding the importance and impact of questions on learning, and deduce its implication for teaching, is to clarify the socio-cognitive mechanisms that are

* E-mail: patriciaalmeida@ua.pt

beneath the process of asking a question. There are many conceptual models trying to explain the questioning process (Lopes, 2013). One of these models was proposed by Hans Van der Meij (1994) and consists of an adaptation of Dillon's Model (1988).

According to Van der Meij (1994) the emergence of a student question implies p*erplexity* (the onset of questioning) and the *formulation of a question* (asking). Each of these two phases might be promoted or blocked, facilitating or hindering the emergence of a question. In the following section we will explore further each phase, complementing it with empirical evidence.

Perplexity

The emergence of a question implies the perception of a phenomenon or information that does not fit into the mental scheme of the potential questioner. This cognitive conflict (Graesser and Person, 1994) might be triggered internally or externally. The majority of studies dedicated to this thematic verified that internal perplexity leads to more in-depth and complex questions (Almeida, 2007, Van Zee, Iwasyk, Kurose, Simpson and Wild, 2001).

But if in one way external stimulus may enhance or promote perplexity, and therefore questioning, the ability of reaching into a perplex mood seems also to be conditioned by the cognitive abilities of the potential questioner (Levin, 2005). There are several investigations that indicate that the reduced number of students' questions during classroom is associated to the inability of detecting inconsistencies (Moreira, 2012). Otero and Graesser (2001) in their PREG Model also approached the relationship between questions frequency and students' cognitive ability, presenting the following hypotheses:

- *Knowledge collision:* more knowledge increases the probability of incompatibilities between the information and his/her own knowledge, increasing uncertainty and therefore the possible number of questions.
- *Knowledge deficiency:* the number of questions should decrease with decreasing knowledge, since the lack of knowledge hinders the identification of gaps in information.

The Formulation of a Question

During this second stage, and through verbal articulation, the learner transforms his/her perplexity into something more tangible, namely ideas and words, which are then organized and structured into a question. Many authors stand along Van der Meij's model arguing that the ability to formulate a question is intimately linked to the verbal abilities of the learner (Chin and Osborne, 2008; Olney, Graesser and Person, 2012).

After mentally articulated, the questions can be answered internally or made 'public' being 'socially expressed' (Chin and Osborne, 2008). Naturally, like the previous stage of perplexity, the formulation of a question also depends of the interaction between individual factors of the students and external factors from the environment. For example: students with low self-esteem tend to report some levels of fear in being teased by colleagues, or even the teacher, since they tend to consider that the question is not relevant enough to justify the interruption of a class (Levin, 2005). On the other hand, students with high self-esteem are particular sensitive to the costs/risk involved in expressing a doubt, frequently deciding not to share the question (Moreira, 2012). Research has not yet been able to evidence clear relationships between the frequency of expressing questions and the abilities and/or personal traits of the students. One possible explanation is that the same characteristic may have contradictory effects considering the stage of questioning (Costa, Caldeira, Gallastegui and Otero, 2000). For example, in a study conducted by Van der Meij (1998), students with low school performance tend to mention during the interview that they have high perplexity levels, but this was not associated to a higher number of questions during classes considering the same students. These results indicate that low level of knowledge promotes perplexity but seems to inhibit the expression of the question, since 'weak' students have frequently low self-esteem perception. Two relevant ideas considering (university) science classes emerge from here:

- the absence of questions from the student can be interpreted as the students "knows all" or the student "does not know anything." Only interaction between students and/or teachers, initiated by the student question, will help to unravel this pedagogical uncertainty;
- the more students ask questions the more he/she contributes to the collaborative knowledge construction in the context of the science class, valuing his/her personal input as a learner.

In this sense, questions are powerful learning and teaching devices which should be promoted at science classes, including at university level (Almeida, 2007; Martinho, Albergaria-Almeida and Teixeira-Dias, 2014; Moreira, 2012; Lopes, 2013), in order to reap out its pedagogical advantages. Table 1 presents a brief systematization of the didactical relevance of questions considering the main learning paradigms.

2. Is There a Relationship between Students' Questions and Learning Styles?

While questions are certainly an indicator of students' learning development, they also allow the establishment of a link between Kolb's theory (Kolb, 1984; Kolb, Boyatzis and Mainemelis, 2001; Mainemelis, Boyatzis and Kolb, 2002) and approaches to learning (Biggs, 1994).

Almeida (2007) defined three categories of questions, according to Kolb's experiential learning theory: acquisition questions, specialization questions and integration questions. Each one of these categories is directly related to the acquisition, specialization and integration phases defined and characterized by Kolb et al. (2001). Students within each of these phases tend to ask questions that are congruent with their stage of development (Figure 1).

Table 1. The role of questions in classrooms considering the main learning and teaching paradigms

Main Learning and Teaching Paradigms	Concept of Learning and the learner			Main Learning Models/ Theories	Didactical implications considering the role of questions
Behaviorism The mind is perceived as a 'black box'. Its activity is only observable through responses that emerge as reaction of stimulus. Learning is conceptualized as a behavior, being extensively structured. Training (repeated practice) and memorization are highly valued.	Learning as an individual and objective task	The individual is mainly passive (low level of processing)		Classical conditioning (Pavlov) Operant conditioning (Skinner)	- Teacher questions as the main didactical tool (instead of student questions); - Teacher questions may be used as tools to lead the students to specific responses/behaviors. Teacher questions may be used to punish or praise the student.
Cognitivism and Socio-Cognitivism The focuses are the mental processes of the students. Being these processes not observable they are inferred indirectly; Recognizes the influence of others (such as the teacher) on the learning process; however learning is individual and happens in the mind of each learner.		The learner has an active role	Medium-High level of information/knowledge processing	Genetic Epistemology (Piaget). Significant Learning (Ausubel). Cognitive Development (Bruner) Social Cognitive Learning Theory (Bandura)	- Teacher questions as thinking tools - teacher questions are used to give students clues/orientations that may help them to organize information and retain knowledge . the teacher uses questions as advanced organizers of information and/or tasks; . the teacher uses a specific sequences of questions to support the students in his/her reasoning (normally general questions first and then, throughout the interaction with the students, more specific questions are used – socratic questioning) - Student questions are indicators/observable evidences of what happens inside the students mind. Questions as products of (re)organization efforts of the mental schemes (it allows the teacher to distinguish what the student already has learned and what is still not learned);
Constructuvism and Socio-constructivism The learning context is highly valued, namely the social interactions with others (teachers, peers and family); Learning is influenced not only by reason but also by emotions (motivations).	Learning as an personal, and also social activity			Experiential Learning (Dewey) Social development (Vygostky)	- Questions as a didactical tool integrated in more global and complex teaching-learning tasks that involve reflection and interaction with others (e.g. case studies, problems) aiming the facilitation of learning

Students in the acquisition phase ask mainly questions that are related to simple facts and concepts – acquisition questions. Students within the middle phase ask mainly specializations questions, those that go beyond the mere search for information. Students in the integration phase show the ability to ask higher-level questions. However, at times, these students also asked acquisition and specialization questions, according to their needs and to the learning environment.

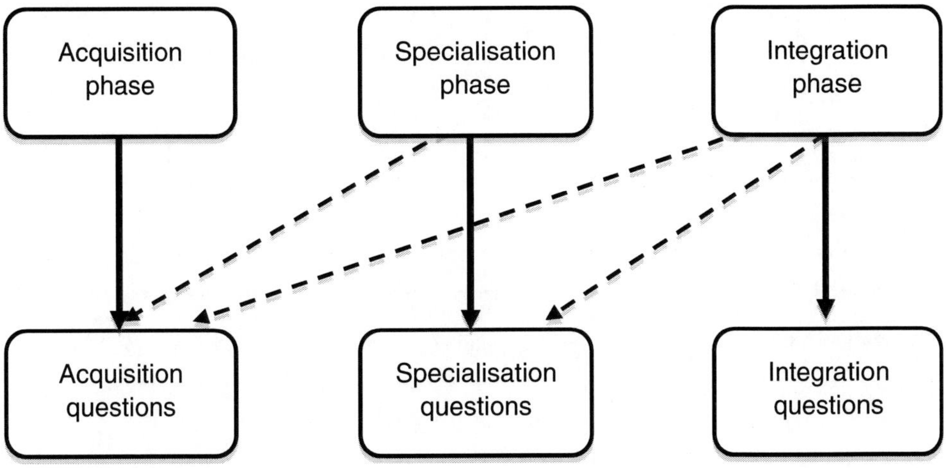

Figure 1. Relationship between students' questions and Kolb's learning theory of growth and development.

In 2000, Chin and Brown established a relationship between students' questions and deep and surface approaches. Almeida, Pedrosa de Jesus and Watts (2011) confirmed this finding, but went a little bit further, suggesting a third approach. When relating students' questions to approaches to learning three categories of questions were proposed: surface, meso and deep questions (Figure 2). Students adopting a surface approach raised mainly surface questions, whose answer require only recall of information. Deep questions showed a clear effort to use knowledge in an innovative way, and revealed reflection and critical thought. These were exclusively raised by students using a deep approach. Students using consistently a deep approach showed the ability to ask different kinds of questions according to the learning environment demands. Thus, besides asking deep questions, these students could also ask surface and meso questions. Students standing in between surface and deep approaches asked mainly meso questions, that go beyond the mere search of information, but did not reach a deep level.

Thus, each one of the classification systems presented has three categories of questions. When analyzing these categories it is possible to identify an obvious congruence between acquisition and surface questions, specialization and meso questions and also between integration and deep questions. Even if defined according to different theories (Kolb's experiential learning theory and approaches to learning), the characteristics of integration and deep questions are clearly equivalent, happening the same to specialization and meso questions, as well as to acquisition and surface questions (Figure 3).

Thus, students' questions represent a powerful tool that allows to establish a relationship between approaches to learning and Kolb's theory (Almeida, 2007; Almeida, et al., 2011) that represent two different dimensions of learning styles (Riding and Rayner, 1998), and that lie in different theoretical backgrounds (Kolb, 1984; Marton and Saljo, 1976).

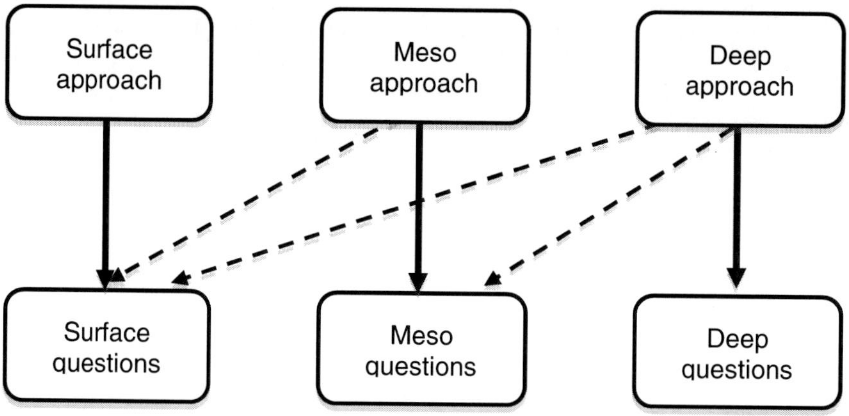

Figure 2. Relationship between students' questions and approaches to learning.

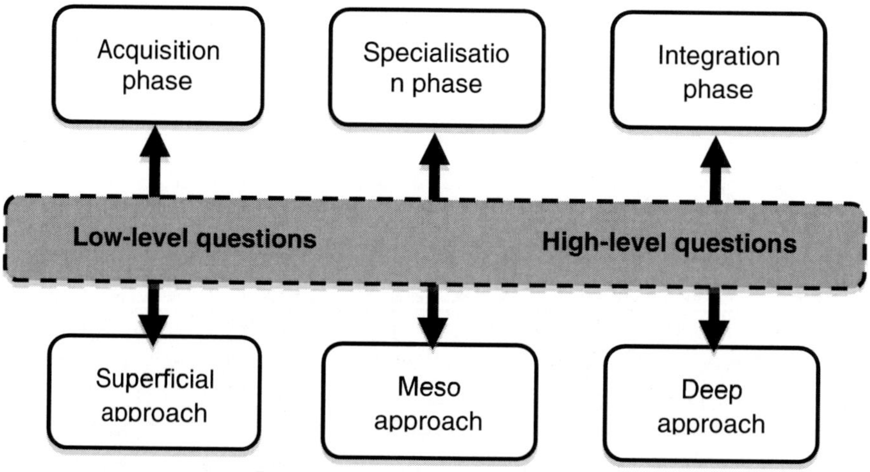

Figure 3. Relationship between students' questions, approaches to learning and Kolb's theory of growth and development.

Nevertheless, both theories have the similar issues in common: that there is a continuum in learning development and students can "move" along this continuum. They might remain at the base – acquisition phase or superficial approach – or move up until they reach a higher-level of thinking, the integration phase or deep approach. Between the base and the top of this continuum, there is learning space where students have "in between" characteristics – the specialization phase or the meso approach. It is important to underline that the transitions between stages does not always occur in a linear manner. Often, recurring oscillations are necessary before a student moves from a lower level to an upper level.

3. Do Questions 'Dress' by Gender?

Concerning student questioning, few studies have focused on gender and even fewer have concentrated on gender differences on Higher Education. However, Wood (2009) states that although women perform better at every levels of education and earn more and higher degrees than males do, women still face biases and barriers in particular fields, namely mathematics and sciences.

Despite the longstanding recognition of the existence of gender differences in verbal communication (Wood, 2009; Tannen, 1990), the few existing studies are not consensual. On the one hand, Pearson, West and Turner (1995) stated that it is not clear which gender raises more questions. On the other hand, Jones, Howe and Rua (2000) observed that boys are less frightened to pose questions than girls. Blum (1999) undertook an investigation to compare the questioning patterns of boys and girls, either in class or online, and concluded that girls ask more questions than boys in class, while boys ask and answer more than girls in online environments.

3.1. Questioning and Gender on a First Year University Chemistry Class

Facing such a controversial issue and considering the great importance of students' questions in the processes of knowledge construction, Martinho, Albergaria-Almeida and Teixeira-Dias (2012a, 2012b, 2012c, 2014) developed a study aiming to investigate and characterize students' questioning profiles according to their gender and to the learning environment in which they are immersed (such as classes or online environments). With this in mind, a number of strategies to increase the opportunities provided for first year chemistry student questioning were implemented.

Findings show that students prefer not to pose questions, which can be understood if we consider students' perception of university classes as impersonal and intimidating (DeBourgh, 2008). Students tend to feel they are passive recipients of the instructor's lecture rather than active participants in a student-instructor interaction (Mayer et al., 2009).

Despite that the importance of student questioning was emphasized from the first class, facing such a different setting from what they were used to, these first year students interacted very little with the lecturer. This shortage of student-teacher interaction is, according to DeBourgh (2008), due to the students' reluctance to express an unpopular opinion and fear to be identified as uninformed or unprepared.

Another finding shows that a very small number of students' questions were written in the pieces of paper provided in class. Most students prefer to pose questions orally to the teacher because that means that they will get an answer immediately. Of all the students that were interviewed, the few who recognized the advantages of written questions alleged that this way it was easier to pose questions to the teacher because of the reduced exposure before the colleagues.

It was also noticed that males pose more questions in class than females. According to Jaasma (1997), male students generally participate more than female students, which could imply that boys would ask more questions to teachers than girls. On the other hand, it could also be due to the fact that on this class male students outnumber female students in class.

Another evident aspect was that by the end of the semester students posed more questions online than during the beginning or middle of the semester, but those questions were mainly referring to the evaluation procedures rather than content. It could be supposed that as the term was reaching an end, students were increasingly getting more worried with the course evaluation and that they were extremely interested in figuring out the best way to have good marks.

Regarding the higher number of questions students ask in chemistry practical laboratory classes in comparison to the questions asked during lectures, it may happened because students had to prepare the lab experiment at home before the class, which meant that they would have time to think, compare, analyze and reflect on the work being developed. Moreover, during the interviews most students of both genders said that they felt more comfortable in posing questions in practical laboratory classes because classes were smaller and they really needed to ask everything they didn't know because they were evaluated in every class. Again, the preoccupation with evaluation was a reason for students' increased questioning.

Table 2. Students' common opinions concerning the effectiveness of the Moodle's forums

Common opinions	Excerpts of students references during the interviews and focus groups
Time for response in classes was faster to get an answer than through the forums	*I think it is easier to pose our questions orally in class, to get an answer immediately, and the issue be solved straight away.*
Receiving an answer orally gives students the possibility to clarify any aspect, which isn't clear at first	*I prefer to get an answer orally, than written on the Moodle, because orally it is more personal. It is easier to understand, if another question emerges immediately after getting a response. It is easier to clarify us.*
Rejection of visibility before the class	*I think that students are influenced by the fact that whatever is written on Moodle is sent to everyone's e-mail. (...) I prefer to send an e-mail directly to the teacher. Maybe there is that unconscious notion of the group's opinion.*
Receiving regular e-mails with every new interaction on Moodle keep students alert	*I would regularly read what appeared in my e-mail account, but I didn't visit the Moodle.* *I was forced to follow the challenges that were being inserted on the Moodle's forum, because they were immediately sent to my e-mail.*
Lacking technical skills to interact with Moodle was rejected by most students	*It is a platform very easy to accede, easy to understand how it works and I believe it had many advantages, such as enabling question posing, download of materials (...)*
If the Moodle was used to post regular activities, students would use it regularly and interact more	*I haven't visited the Chemistry Moodle as many times as I did for other courses, because other courses have all the material on Moodle. (...) Regarding Chemistry, the material that we need is included in the book and in the guidebook* [delivered to students in printed format at the beginning of the semester].
Moodle is an advantage for students to get a succinct answer	*On Moodle the answers we get are perhaps more succinct than those we get orally in class, because the answers in class (...) usually include a lot of information and we end up loosing the content of the question that had been initially asked.*
Online participation not being considered for assessment	*If it* [participation on Moodle] *was considered for evaluation (...), the number of questions would increase dramatically.*

Finally, students asked more questions in any kind of class than through the online tool (Moodle) provided during the course. In fact, when interviewed, students indicated that they preferred traditional lectures or chemistry practical laboratory classes as a questioning learning environment, rather than one involving the content management system used during the course, the Moodle. It could be that the content management system available wasn't friendly enough to increase students questioning (Martinho, Albergaria-Almeida and Teixeira-Dias, 2014). Furthermore, as a content management system it is primarily purposed at the "storage" of content, rather than at the support of interactions. During the interviews and focus groups most students didn't recognize the advantages of question-posing through Moodle and alleged that it would take more time for them to have an answer or an explanation than in class (Table 2).

Conclusion

Student-generated questions play a significant role in motivating meaningful learning and can serve different functions within this. These functions can include confirmation of expectations, answers to unexpected puzzles, and filling a recognized knowledge gap. The questions that learners ask are also indicative of their need for resolution in their thinking, for understanding within the domains in which they are working and studying, and for some degree of interaction with both teachers and other students within sessions. Student questioning, particularly at the higher cognitive levels, is also an essential aspect of problem solving. Besides helping students learn, student questioning can also guide teachers in their work. Some researchers have explored the potential for using students' questions to influence the curriculum. Some questions indicate that students have been thinking about the ideas presented and have been trying to extend and link these with other things they already know. Questions can also reveal much about the quality of students' thinking and conceptual understanding, their alternative frameworks and confusion about various concepts, their reasoning and what it is they want to know . Therefore the invitation to 'Ask a question!' might be one step further considering the improvement of learning (and teaching) at university science classes.

Acknowledgment

We acknowledge the financial support of the Research Centre for Didactics and Technology in Teacher Education (CIDTFF), University of Aveiro, Portugal, and of the Portuguese Foundation of Science and Technology (FCT; SFRH/BD/74511/2010).

References

Almeida. P. (2007). Questões dos estudantes e estilos de aprendizagem – um estudo com um público de Ciências no ensino universitário (Students' questions and learning styles – a study with Science University students). PhD thesis. Aveiro, Portugal: University of

Aveiro. Online available: http://oatd.org/oatd/ record?record=handle%5C: 10773%5C% 2F1461

Almeida, P., Pedrosa de Jesus, H. and Watts, M. (2011). Kolb's learning styles and approaches to learning through the use of students' critical questions. In S. Rayner and E. Cools (Eds.), International Perspectives On Style Differences In Human Performance: Leading Edge Research, Theory And Practice (pp. 115-128). London: Routledge.

Biggs, J. (1994). Approaches to learning: nature and measurement of. In T. Husen and T. N. Postlethwaite (Eds), The International Encyclopaedia of Education, Vol. 1, 2nd ed., Pergamon, Oxford.

Blum, K. (1999). Gender differences in asynchronous learning in higher education: Learning styles, participation barriers and communication patters. *Journal of Asynchronous Learning Networks,* 3 (1), 44-66.

Chin, C. and Brown, D. E. (2000), Learning in Science: A Comparison of Deep and Surface Approaches. *Journal of Research in Science Teaching,* 37, 109–138.

Chin, C. and Osborne, J. (2008). Students' questions: a potential resource for teaching and learning science. *Studies in Science Education*, 44(1), 1-39.

Costa, J., Caldeira, M. H., Gallastegui, A., and Otero, J. (2000). An analysis of question asking on scientific texts explaining natural phenomena. *Journal of Research in Science Teaching,* 37, 602-614.

DeBourgh, G. A. (2008). Use of classroom "clickers" to promote acquisition of advanced reasoning skills. *Nurse Education in Practice*, 8 (2), 76-87.

Dillon, J. T. (1988). The remedial status of students questioning. *Journal Curriculum Studies,* 20(3), 197–210.

Graesser, A. C. and Person, N. K. (1992). Question asking during tutoring. *American Educational Research Journal,* 31, 104-137.

Graesser, A. C. and Olde, B. A. (2003). How does one know whether a person understands a device? The quality of the questions the person asks when the device breaks down. *Journal of Educational Psychology*, 95, 524-536.

Jaasma, M. A. (1997). Classroom communication apprehension: Does being male or female make a difference?, *Communication Reports*, 10 (2), 219-228.

Jones, M. G., Howe, A. and Rua, M. J. (2000). Gender differences in students' experiences, interests, and attitudes toward science and scientists. *Science Education*, 84(2), 180-192.

Kolb, D. A. (1984). Experiential Learning: Experience as a Source of Learning and Development, Prentice-Hall, Englewood Cliffs, NJ.

Kolb, D. A., Boyatzis, R. and Mainemelis, C. (2001). Experiential learning theory: previous research and new directions. In R. J. Sternberg and L. Zhang (Eds.), Perspectives on thinking, learning and cognitive styles (pp. 227-247). London: Erlbaum Associates.

Levin, A. (2005). Lernen durch fragen: wirkung von strukturierenden hilfen auf das generireren von studienfragen als begleitende lernstrategie. PHD thesis. Berlin: Technischen Universität Berlin.

Lopes, B. (2013). Abordagens ao Ensino e Práticas de Questionamento no Ensino Superior (Teaching Approaches and Questionning Pratices in Higher Education). PhD thesis. Aveiro, Portugal: University of Aveiro. Online available: https://www.researchgate.net/ profile/Betina_Lopes/publications?pubType=thesisandev=prf_pubs_the

Mainemelis, C., Boyatzis, R. and Kolb, D.A. (2002). Learning styles and adaptive flexibility: testing the experiential theory of development, *Management Learning*, 33(1), pp. 5-53.

Martinho, M., Albergaria Almeida, P. and Teixeira-Dias, J. (2012a). Student questioning in higher education: Does gender make a difference? *Cypriot Journal of Educational Sciences,* 7(4), 334-347.

Martinho, M., Albergaria Almeida, P. and Teixeira-Dias, J. (2012b). A gender perspective on student questioning upon the transition to higher education. In N. Popov, C. Wolhuter, B. Leutwyler, G. Hilton, J. Ogunleye and P. Albergaria Almeida (Eds.), International perspectives on education (pp. 418-424). Sofia: Bulgarian Comparative Education Society.

Martinho, M., Albergaria Almeida, P. and Teixeira-Dias, J. (2012c). Students' questions in Higher Education chemistry classes according to their gender. *Procedia – Social and Behavioral Sciences,* 47, 835-840.

Martinho, M., Albergaria Almeida, P. and Teixeira-Dias, J. (2014). Do Moodle's forums foster student questioning? The students' voice. In A. D. Benson and A. Whitworth (Eds.), Research on Course Management Systems in Higher Education (pp. 119-142). Information Age Publishing.

Marton, F. and Saljo, R. (1976). On qualitative differences in learning: I. *Outcome and process, British Journal of Educational Psychology,* 46, 4-11.

Mayer, R. E., Stull, A., DeLeeuw, K., Almeroth, K., Bimber, B., Chun, D., ... Zhang, H. (2009). Clickers in college classrooms: Fostering learning with questioning methods in large lecture classes. *Contemporary Educational Psychology*, 34 (1), 51-57.

Moreira, A. (2012). O questionamento no alinhamento do ensino, aprendizagem e avaliação. (The role of questioning in the alignment of teaching, learning and assessment). University of Aveiro: Aveiro.

Olney, A. M., Graesser, A. C. and Person, N. K. (2012). Question generation from concept maps. *Dialogue and Discourse*, 3(2), 75 - 99.

Otero, J. and Graesser, A. C. (2001). PREG: Elements of a model of question asking. *Cognition and Instruction*, 19(2), 143-175.

Pearson, J., West, R., and Turner, L. (1995). Gender and communication (3rd ed.). Dubuque: Brown and Benchmark Publishers.

Riding, R. J. and Rayner, S. (1998). Cognitive Styles and Learning Strategies. London: David Fulton Publishers.

Tannen, D. (1990). You just don't understand: Women and men in conversation. New York: Ballantine.

Van der Meij, H. (1994). Students questioning a componential analysis. *Learning and Individual Differences,* 6(2), 137-161.

Van der Meij, H. (1998). The great divide between teacher and student questioning. In S. A. Karabenick (Ed.), Strategic Help Seeking: Implications for Learning and Teaching (pp. 195-218). New Jersey: Lawrence Erlbaum Associates.

Van Zee, E. H., Iwasyk, M., Kurose, A., Simpson, D. and Wild, J. (2001). Students and teacher questioning during conversations about science. *Journal of Research in Science Teaching,* 38(2), 159-190.

Wood, J. T. (2009). Gendered lives: Communication, gender, and culture (8th ed.). Belmont: Wadsworth.

In: New Developments in Science Education Research
Editor: Nathan L. Yates
ISBN: 978-1-63463-914-9

Chapter 5

SCAFFOLDING INQUIRY-BASED SCIENCE AND CHEMISTRY EDUCATION IN INCLUSIVE CLASSROOMS

Simone Abels*

University of Vienna,
Austrian Educational Competence Centre Chemistry, Vienna, Austria

ABSTRACT

Since the ratification of the UN Convention on Rights of Persons with Disabilities inclusion is on the agenda. In science education inquiry-based learning is, among others, recommended to deal with the diversity of students in a classroom. However, teachers struggle with the implementation of inquiry-based science education and view heterogeneity rather as a burden than an asset.

Therefore, this research project is dedicated to the teachers' scaffolding teaching science in inclusive classes. Specifically, within the framework of a qualitative case study, the focus lies upon inquiry-based learning environments with different levels of openness and structuring and how they can be orchestrated to deal with diverse learning needs of all students in one classroom. The research study is conducted in cooperation with an inclusive urban middle school (grade 5 to 8), where students with and without special educational needs learn together.

Initially, the article introduces the inclusive school with its specific way and challenges – not only the science classes – as inclusion is holistic school development. Subsequently, two science formats are particularly focused and contrasted: guided inquiry-based chemistry lessons as well as the format *Lernwerkstatt* (German for workshop center), an open inquiry setting. The aim of the *Lernwerkstatt* is that students work self-dependently on their own scientific questions innervated by carefully arranged phenomena and materials.

Using participant observation and videography data were collected in two 8th grade chemistry classes and nine *Lernwerkstatt* classes. Additionally, audio recordings of informal conversations with the science teachers are used for analysis, which follows the rules of Grounded Theory by Kathy Charmaz. The systematics of Grounded Theory was

* Email: Simonc.abcls@univic.ac.at

chosen as it focuses on social phenomena and processes, e.g., teaching processes which need to be analyzed to outline inclusive science education. The differences in the teachers' scaffolding of the open versus the guided setting will be explained.

INTRODUCTION

Inclusion is defined as "**a process** of addressing and responding to the diversity of needs of all learners through increasing participation in learning, cultures and communities, and reducing exclusion within and from education. It involves changes and modifications in content, approaches, structures and strategies, with a common vision which covers all children of the appropriate age range and a conviction that it is the responsibility of the regular system to educate all children" (UNESCO, 2005, p. 13, original emph.).

This means, inclusion focuses on the diversity of all students and regards their diversity as an asset for learning (Sliwka, 2010). Students can differ in many dimensions, e.g. gender, age, ethnicity, culture, socio-economic background, mental and physical abilities and many more (Krell, Riedmüller, Sieben & Vinz, 2007). These dimensions can affect language skills, interests, prior knowledge, experiences, social skills etc. and are thus present in learning contexts. The question is how teachers can successively make the diversity productive for learning and the development of the students.

Since the almost worldwide ratification of the UN Convention on Rights of Persons with Disabilities (United Nations, 2006), inclusive schools are politically demanded and mandatory. To raise the achievement of all students is an "ethical imperative" (European Agency for Development in Special Needs Education, 2012, p. 4). However, the challenge is to implement the convention locally. Schools are often not provided with appropriate resources and many teachers do not feel educated enough to teach inclusively, especially not subject teachers in secondary schools. They rather feel overcharged to cope with individual differences in their classrooms (cp. Markic & Abels, 2014).

The transfer of inclusive teaching principles to subject teaching in secondary schools seems to be an enormous challenge. One reason is that inclusion is a profound endeavor; it requires holistic school development, collaboration among colleagues and with policy stakeholders, evaluation and reflection to ensure an ongoing development, visionary school leaders, a support system, professional development etc. (European Agency for Development in Special Needs Education, 2012). The single subject teacher is only one, thereby very important component contributing to successful inclusion. Another reason is that inclusion is linked to normative ideals demanding positive attitudes of teachers towards diversity (Avramidis & Norwich, 2002). It is understandably difficult to maintain a positive attitude if political decisions, the provision of resources, societal demands or the quality of teacher education contradict inclusive school systems. Teachers who, nevertheless, try to meet the demands of inclusion have to be praised.

Models of effective inclusive subject teaching and appropriate ways of professional development need to be found and evaluated that maintain the ethical imperative, but realistically take the local school systems and their conditions into account. The application of factors which have shown to be effective in inclusive teaching (Meijer, 2010), for example, cooperative teaching and a clear classroom management, peer tutoring among students and heterogeneous grouping as well as problem-based learning strategies with increasing students' responsibility, seems to be challenging (Stroh, 2014). The specific practice of inclusion at classroom level is not well elaborated (Florian & Black-Hawkins, 2011). The culture of certain school subjects almost appears to be incompatible with inclusive principles; science seems to be such a subject. This will be elaborated in the next paragraph before a case study is introduced, where science education at a school striving to be inclusive is researched qualitatively.

SCIENCE EDUCATION AND INCLUSIVE EDUCATION – TWO INCOMPATIBLE CULTURES?

Science can be thrilling, fascinating, intriguing and challenging. There are science teachers, science teacher educators and science education researchers who work intensively on maintaining these aspects in school science. The development of science education programs perceived as relevant and motivating is on the agenda. But still, school science counts as boring, authoritarian, theoretical, difficult and deterrent (cp. Sjøberg & Schreiner, 2010). Why is that? School science mostly in western countries fails to show its relevance for the individuals, potential careers and the society they live in (Stuckey, Hofstein, Mamlok-Naaman & Eilks, 2013). An expert group of the European Commission concludes that the way science is taught is a reason for the decline of interest among students (Rocard et al., 2007). From the TIMSS video study is known that in German physics classes, for example, a teacher-led IRE (initiation – response – evaluation) discourse is a dominant procedure (Seidel et al., 2006), which excludes most students from participation. Also reading and writing scientific texts are common tasks in science classes, but do not allow for equal learning opportunities (Scruggs & Mastropieri, 2007). "These whole-class approaches usually do not provide the possibility to work in one's own speed and on one's own level of difficulty" (Markic & Abels, 2014, p. 275). Furthermore, science lessons are often content-focused instead of skill-led (Roth, 2009), but the content is not presented as person-related and does not become meaningful. Rather an overload of isolated facts is taught that are difficult to transfer to everyday life contexts (Gilbert, 2006). This pattern is repeated and rarely reflected in the pre-service education of teachers (Wildt, 2003).

Science education in the traditional way just described has to be considered as incompatible with inclusive practice as it cultivates exclusion mechanisms and is rather oriented on some students who will maybe pursue a career in science or technology. On the contrary, inclusive practice at classroom level demands learning opportunities and settings that are "sufficiently made available for everyone" facilitating participation for all ("inclusive pedagogical approach") instead of using settings that are appropriate for most learners with additional differentiation for some ("additional needs approach") (Florian & Black-Hawkins, 2011). Usually in traditional science education, neither the former nor the latter is applied. A big challenge is to resist from or change determinist thinking that underpins labelling, categorizing and excluding (Hart, Drummond & McIntyre, 2007).

The inclusive pedagogical approach is more in line with the UN convention than the additional needs approach. It even seems to be more suitable and better applicable for the situation of classroom/subject teachers who have not been educated to teach students with special or additional needs. This means, instead of diagnosing learning needs and developing differentiated measures for some students with the consequence of ability-labelling and stigmatization, approaches have to be chosen or enriched that allow learning for all. Researchers need to provide convincing evidence these choices have to be based on. Higgins, Kokotsaki and Coe (2011) emphasize effective feedback, meta-cognition, self-regulated strategies and early intervention as high impact approaches. According to Florian and Black-Hawkins (2011, p. 821) the inclusive pedagogical approach includes that:

- "Students choose how, where, when and with whom they learn
- Teachers create options and consult with each student about how they can help
- Teachers create the conditions that support students to work with different groups
- Classroom teacher consults with colleagues including those in learning support to share ideas about teaching and learning
- Students are trusted to make good decisions about their learning."

Feyerer (2007) mentions a range of teaching principles that have shown to foster inclusive practice (figure 1). Extending the right hand side means to approach inclusive teaching.

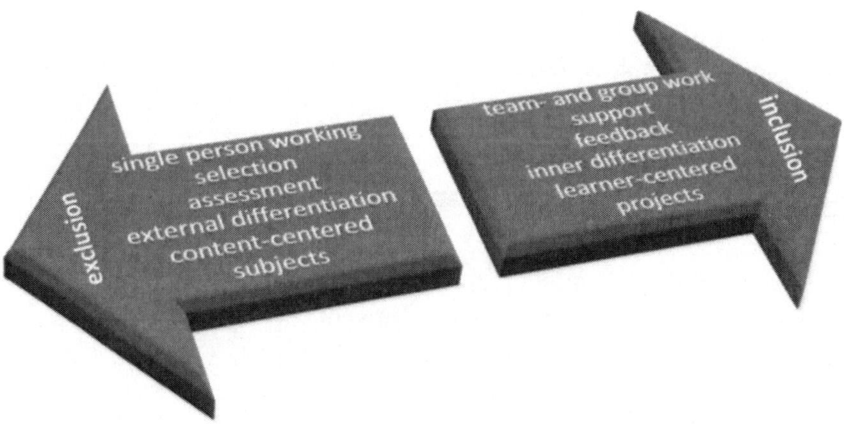

Figure 1. A range of teaching principles (cp. Markic & Abels, 2014).

Moreover, activity-oriented, experience-based, interdisciplinary, cooperative, interactive, flexible learning settings based on a variety of teaching methods and joint relevant content allow for a high amount of self-determination and participation (UNESCO, 2005). These principles have to be 'translated' for and applied in classroom practice.

Also in science education some approaches have the potential to be inclusive pedagogical approaches. Inquiry-based learning counts as one of these approaches. It is recommended to foster students across the ability range and has shown to be effective on primary and secondary level, with girls and boys as well as for low and high achievers (Rocard et al., 2007). But teachers struggle with the implementation especially in inclusive classrooms. Certain conditions have to be considered when implementing inquiry-based science education.

Inquiry-Based Learning to Reach Compatibility between Science and Inclusion

Inquiry-based science education involves all the activities where students come to an understanding of scientific concepts and the work of scientists (National Research Council, 2000). Activities are, among others, posing questions, hypothesizing, collecting data, observing, analyzing data, interpreting, presenting and discussing results as well as making evidence-based conclusions. The activities do not follow each other linearly; there can be a lot of different ways to approach and conduct an inquiry. Using this approach three goals are aimed at: to learn scientific content, to learn to do inquiry and to learn about inquiry (Abrams, Southerland & Evans, 2008). Cooperation and communication are essential features of inquiry-based settings. Depending on their prior knowledge, their experience with this kind of learning, social skills and abilities in self-regulated learning, students need more or less support and scaffolding to conduct an inquiry (Abels, accepted).

Often forgotten is, that opportunities have to be created to get used to a new learning culture and to acquire the necessary skills to do inquiry and reflect inquiry processes. Therefore, the implementation of the inquiry approach has to happen stepwise with a successive enrichment of skills. A helpful tool is provided by the levels of inquiry (table 1). "Instruction should gradually and systematically move from Level '0' activities with the ultimate goal being some Level '3' activities" (Lederman, Southerland & Akerson, 2008, p. 32).

Table 1. The levels of inquiry (Blanchard et al., 2010, p. 581)

	Source of the question	Data collection methods	Interpretation of results
Level 0: verification	Given by teacher	Given by teacher	Given by teacher
Level 1: structured	Given by teacher	Given by teacher	*Open to student*
Level 2: guided	Given by teacher	*Open to student*	*Open to student*
Level 3: open	*Open to student*	*Open to student*	*Open to student*

With each level the responsibility and self-dependent activity of the students are increasing while at the same time the direct instruction of the teacher is decreasing. On a level 0 inquiry instructions are given for every step of the inquiry process, even the results and their interpretation are given. The challenge for the students is to replicate the phenomenon and manage the handling of the tools. Teachers can use this level to introduce a new topic or new devices, to promote manual skills, to foster the comprehension of written instructions, to implement prior knowledge or to reduce potential risks. Thus, level 0 is not only appropriate for low achievers as earlier studies claimed (cp. Ellis, 1993; Werning & Lütje-Klose, 2007); its application depends on the learning goals and the context. Level 0 fosters certain steps of an inquiry process and should not be left out. As the opening of the inquiry process begins at the back, on level 1 the interpretation of results is done by the students themselves. Level 1 is seen as the typical student experiments. The students can learn, for example, "to document observations and interpret them in the team, apply knowledge to come to conclusions and judgments, to justify conclusions with evidence-based arguments or to present and discuss results" (Abels, 2014b, p. 135). On level 2 the choice and application of methods is additionally in students' hands while the teacher is providing the question(s) whereas on level 3 the students work on their own questions of interest and take responsibility for the whole investigation process. As the investigations of the groups can develop in very different directions, teachers are challenged to structure classroom practice flexibly. Provided materials can reduce or increase the number of options to inquire a question. Posing high-level-type questions that can be researched in school, i.e. level 3, counts as most difficult for students (Hofstein, Navon, Kipnis & Mamlok-Naaman, 2005). Teachers need to reflect with the students how to phrase a question which is researchable with the school's resources in the given time frame. It is easier to carry out a level 3 inquiry when level 1 or 2 investigations have taken place before and follow-up questions in the same topic are developed. Martin-Hansen (2002) calls this "coupled inquiry".

Not for every student level 3 is the optimal level to be achieved (Blanchard et al., 2010). It is recommended to choose the level in accordance with the prior knowledge, skills and experience of the students. Especially for students with learning difficulties level 2 is recommended as it allows for a balance of openness and structuring (Scruggs, Mastropieri & Okolo, 2008). To make an appropriate choice teachers also have to take into account the available resources, e.g. personnel support, materials, facilities etc. Especially the transition from one level to another has to be structured carefully. Each new skill, e.g. planning and carrying out investigations, analyzing data or engaging in argument from evidence should be fostered predominantly (Achieve Inc., 2013). However, the increase of levels of inquiry does not necessarily implicate an increase of difficulty. For some students level 0 can be more puzzling than the creation of questions on level 3. Each level has its own challenges and foci, also for teachers scaffolding the process. Each level requires the teacher to be in a different role.

The level-based inquiry approach becomes inclusive when different students get the opportunity to work on different levels in the same classroom on the same topic at the same time. This is possible by providing supportive materials and individualized scaffolding. Starting a level 2 inquiry, for example, means that the teacher phrases the question to be investigated. The students plan the inquiry in small carefully selected groups. Teams that struggle with the planning process could draw on supportive material, e.g. a prepared material table, helping cards with hints on the devices to be used or stated hypotheses, a sketch of an experiment setup or a joker to spy on the ideas of another group. The more the process is structured the closer it gets to a level 1 inquiry. Nevertheless, the students work on the same topic in their zone of proximal development (Vygotsky, 1978). This kind of learning culture needs to be implemented carefully (Ahlring, 2000). It has to be normal and assessed positively to look for support. This is a significant skill which students and also teachers need to develop.

The case study of Sullivan Palincsar, Magnusson, Collins and Cutter (2001) contributes to understand the learning opportunities students get when participating in a guided inquiry-based setting. All students, also those with special needs, made significant learning gains when scaffolded by teachers with advanced strategies, i.e., "(a) monitoring and facilitating student thinking, (b) supporting print literacy, and (c) improving working in groups" (Sullivan Palincsar et al., 2001, p. 24). "When instruction is appropriately presented and modified, students with learning disabilities are very successful at mastering science content"

(Brigham, Scruggs & Mastropieri, 2011, p. 230). The point is that the system has to allow for differentiated learning goals which is not always the case.

An important tool if not the most important one to make an inquiry-based task inclusive is scaffolding. Scaffolding can be understood as careful guidance to help students solve scientific problems which are located in their zone of proximal development (Furtak, 2008). The teachers support the student to reach the next development level. Helpful strategies are, for example, "giving approval, probing learner's ideas, structuring task activities, and providing general hints or specific suggestions that will help the learner throughout the task. Asking questions to the learner and using appropriate (...) materials are other important scaffolding tools" (van der Valk & de Jong, 2009, p. 832). Van Zee and Minstrell (1997) emphasize the "reflective toss" as one way of using questions, whereby teachers give the students responsibility for their thinking and initiate deeper thinking processes when questions follow students' statements. There are more strategies of scaffolding which can support inquiry learning. The practices are listed in the following cited from (Abels, 2014b, p. 135f). Further reading advice is given.

- Teaching mnemonic strategies is effective as students can recall vocabulary and thus have more capacity to learn science concepts (Scruggs & Mastropieri, 2000; Scruggs et al., 2008; Therrien, Taylor, Hosp, Kaldenberg & Gorsh, 2011).
- Spooner, Knight, Browder and Smith (2012) identified task analytic instruction with systematic prompting and feedback as well as time delay as evidence-based practices to support students with disabilities (cp. also Browder et al., 2012).
- Graphic organizers "improve the factual comprehension and vocabulary knowledge of intermediate and secondary students with LD [learning disability] in science" (Dexter, Park & Hughes, 2011, p. 210). They also facilitate longer maintenance of scientific knowledge (ibid.).
- Peer-tutoring has shown to be very successful in supporting students with cognitive disorders (Jimenez, Browder, Spooner & Dibiase, 2012; Scruggs & Mastropieri, 2007).
- Text enhancements, vocabulary learning and other language strategies support a diverse student group in comprehending a science concept and conducting an inquiry (Bakken, Mastropieri & Scruggs, 1997; Markic & Abels, 2013; Mason & Hedin, 2011). Word and picture symbol cards were also shown to be supportive (Browder et al., 2012)
- Targeted questioning by teachers or peers helps students to draw inferences and come to higher levels of comprehension compared to just providing them the knowledge (Mastropieri, Scruggs & Butcher, 1997).
- Differentiated materials enable students of different achievement levels to work on the same topic (Abels & Markic, 2013; Tobin & Tippett, 2013).

Teachers need to clarify the intended development zone of the students and choose the quantity and quality of scaffolding accordingly, sometimes before the lesson, sometimes spontaneously during classroom action. Scaffolding puts teachers in a different role than traditional instruction. They have to endure uncertainty and lack of knowledge to a certain extent. Students have to learn that their teachers cannot be omniscient, but support them in acquiring learning strategies. Pre- and in-service education has to provide more learning opportunities for teachers in this area so that they can get used to different roles and expectations (van der Valk & de Jong, 2009).

In the next chapter a school is described which uses different levels of inquiry-based learning as an inclusive approach to deal productively with the diversity in the classrooms to illustrate how the described approaches can be put into practice. Subject teaching will be contrasted with an interdisciplinary setting.

Implementation of Inquiry-Based Learning in an Inclusive School – A Case Study

Before the implementation is explained, a general description of the school is given. The school cooperating in the research project at hand is an urban inclusive middle school with ten classes (grade 5 to 8) in Austria, where students with and without special needs learn together in the same classrooms. Focal points of support of the special needs students are mental development, learning, emotional and social development as well as language. The school used to be a special needs school, i.e., the school took the inverted way of inclusion (cp. Abels, 2014a). Technically speaking, the school is an *integrative*, not an *inclusive* school, because the Austrian system still demands the official labelling of the special needs students, resources are not provided sufficiently, general teacher education is not adapted to the demands of inclusion and there are still separate school types where students are sent after fourth grade. The latter means that at a middle school usually the high achieving students are missing.

Although the school system does not allow for real inclusion, a holistic approach of school development is undertaken in this case and the teachers constantly strive for more and more inclusive teaching approaches. For each age group a steady team of teachers is responsible to ensure reliable teacher-student-relationships. Educators are special needs, middle school or secondary school teachers working in a team. An internal school development team is responsible for establishing inclusive measures and their evaluation. Measures are, for example, to refrain from grading using formative skill-led assessment instead, to involve the support of a social worker, to offer several coaching programs to support students in learning how to learn, in reading and writing or in motor skills (cp. Minnerop-Haeler, 2013). Teaching in the classroom is supposed to be progressive, i.e., project-oriented, learner-centered, skill-focused, differentiated and problem-based with an emphasis on increasing autonomy of the students.

In science classes these teaching principles are applied by implementing different levels of inquiry-based learning. Two science formats are particularly focused and contrasted in this case study: structured and guided inquiry-based chemistry lessons (because of the author's area of expertise) as well as an open inquiry setting called *Lernwerkstatt*[1].

Between the two formats are systemic differences: Only eighth graders have chemistry. In this school it was decided that students with the focal point of support in mental development, i.e. mentally retarded students, do not participate in the chemistry lessons.

[1] The concept *Lernwerkstatt* is mainly based on the New York workshop centre developed by Lillian Weber (1977). As no appropriate translation exists, the term *Lernwerkstatt* will be used here.

The reasons are the complexity, abstractness and potential risks of chemistry. All the other students, also students with special needs in learning, behavior or language, participate in chemistry lessons. One class is always split in semigroups, so that nine to ten students form one group. Each group receives one hour of chemistry per week. Parallel, the other semigroup has computer science. Sometimes the chemistry teacher asks for a double period with one semigroup. Mainly, IRE discourse, structured and guided inquiry settings are applied.

Lernwerkstatt is organized differently. Every class has one *Lernwerkstatt* per year from grade 5 to grade 8; each lasts three days and every student participates. The aim of the *Lernwerkstatt* is that all students work self-dependently on their own scientific questions innervated by carefully arranged phenomena and materials among one major topic, e.g., light and color, water or small life big time (insects) (figure 2).

Figure 2. Scenery with materials in the Lernwerkstatt "Small life big-time".

"A Lernwerkstatt is described as a room where learners encounter stimulating phenomena, objects and materials which are supposed to trigger questions in their own field of interest (…) to start immediately with an inquiry" (Puddu, Keller & Lembens, 2012, p. 154). That is why *Lernwerkstatt* is a level 3 inquiry. As the students come up with questions, *Lernwerkstatt* cannot be reduced to one science subject, but is an interdisciplinary endeavor. The setting is accompanied by coaches who scaffold the students' learning processes (Hagstedt, 2004; Zocher, 2000). Besides two *Lernwerkstatt* teachers (one of them is the chemistry teacher), coaches are the classroom teachers who are released as much as possible from other duty for the three days, higher education students or assistant teachers. These additional material and non-material resources are organized by the *Lernwerkstatt* teachers themselves. Every *Lernwerkstatt* ends with a festivity where students present findings of their investigation.

These two different settings, a subject-specific and an interdisciplinary one, which differ in the degree of inclusiveness, function to compare the systemic influences on scaffolding in diverse classes. Moreover, within the framework of a qualitative case study, is researched how inquiry-based learning environments on different levels of openness and structuring can be orchestrated to deal with diverse learning needs of all students in one classroom.

Methodological Framework

Data collection took place in two 8th grade chemistry classes and in nine classes who had three days of *Lernwerkstatt* each in the school year 2013/14. The data collection is based on videography, participant observation and informal conversations with the teachers. The chemistry teacher was also teaching in the *Lernwerkstatt* plus another science teacher. Both teachers wore audio recorders to ensure good audio quality. Informal conversations with the researcher, who was an active observing participant in both settings, took place and were also audiotaped. For the analysis presented in this chapter, video scenes were chosen where scaffolding processes could be identified. Video was found to be a powerful, complex, enduring, authentic and flexible tool in education research that allows for deep analysis from multiple viewpoints (Goldman, 2007; Janík, Seidel & Najvar, 2009). Videography facilitates to observe authentic practice and to identify patterns of instruction.

Collected data were analyzed with technical support of the qualitative data analysis program ATLAS.ti (version 7.5) by means of Grounded Theory, following the modes of initial, focused and theoretical coding (Charmaz, 2006). This allows for deep and detailed insights into scaffolding considering it as a social practice. In Grounded Theory it is important that the researcher makes her assumptions transparent and reflects how they affect the interpretations (Breuer, 2010). This is regarded as important as the interpretation of effective teaching can be a personally and theoretically biased endeavor. "Adopting [this] self-reflexive, inductive (or abductive) orientation, theoretical models [get] developed from the observed everyday social phenomena and their subjective representations (for example in the form of narrations)" (Breuer, 2011, para. 2). In line with this approach, the methodological device of contrasting cases is essential. Each level of inquiry-based settings that can be observed in the two science formats (level 1, 2 and 3) is demarcated as one case to be contrasted with each other. A "categorical system with a certain level of abstraction" has to emerge (Breuer, 2011, para. 2). This way of qualitative data analysis based on audio and video recordings has the potential to provide deep insights into social situations and to explain behavior or communication patterns that were not evident before (Griffiths, 2013; Nilsson, 2012). Furthermore, Grounded Theory is seen as a social-scientific methodology that can be appropriately transferred to (science) education questions (Sander, 2008).

RESULTS

Scaffolding processes of the two teachers were coded by an open coding procedure and some major categories (theoretical codings) emerged; two of them are chosen to structure this chapter: scaffolding self-dependent learning and the sway of subject-matter knowledge on scaffolding. Extracts from informal conversations are added to illustrate the data.

Scaffolding Self-Dependent Learning

Scaffolding *Lernwerkstatt* or chemistry lessons can be distinguished by the degree of students' self-dependence. In *Lernwerkstatt* students decide about the question they want to work on, the methods, the materials, the social form (single or teamwork), how intensively they step into the topic, i.e. the level of difficulty, the learning rate and with whom and where they install their place of work. The coaches scaffold the students' learning processes accordingly and have to adapt, often spontaneously, to the ideas, choices and prior knowledge of the students. The challenge is to decide how much input is needed when to maintain the high degree of self-dependence and to facilitate a sense of achievement. If students ask questions, teachers do not just give an answer, but help students searching it by applying scaffolding strategies named above, e.g. the reflective toss. Because of the variety of investigations, it can also happen that the teachers do not know anything about the chosen topic. But they provide assistance in structuring the inquiry process, organizing material, finding ways of gathering information, eventually solving conflicts etc. Then it becomes a joint learning process which is actually experienced as beneficial for both sides, teachers and students.

Analyzing the scaffolding of the *Lernwerkstatt* it is noticeable that some coaching is rather focused on structuring the inquiry process while some coaching seems to be rather concerned with structuring group work, i.e. discipline issues as the following video transcript shows.

> Teacher: "Well, no no no. My problem is that you two are distracted very easily. I want you to sit over there (points to the other side of the room). You are distracted way too easily. That doesn't work."[2]

The teacher uses short sentences and clear, distinct statements. She directs the students so that they have better conditions to work. In comparison, scaffolding the investigation process appears to be different. In the example given below, the students intend to understand how a kaleidoscope is constructed.

> Teacher: So, you drew the outside of the kaleidoscope, good. What is the next step?
> Student: We must think about which stones come into the kaleidoscope.
> Teacher: That is the most important question. What kind of stones do I put inside? How do the stones look like? Look, you have so much stuff there, just try it.

The teacher summarizes the students' work done so fare and leads their plan a step further. She strengthens important points and rephrases the students' statements as questions. Additionally, her utterances have an open, inviting, motivating character instead of being distinct. As it is a very challenging phase the most effort in scaffolding *Lernwerkstatt* is done so that every student finds a question. Teachers need to be aware which students need support here. It sometimes occurs especially with special needs students that they do not find a question, but work on a topic like "the color blue". The teachers in this project take this issue in hand and work on ideas to scaffold the phase of finding questions differently (cp. Abels, 2014b), for example, by providing stations first which lead to follow-up questions (coupled inquiry). In chemistry, on the contrary, the most effort in scaffolding is done so that every student finds an answer. In chemistry lessons students cannot decide independently about the topic or question they want to work on. The content is given as well as the social form, the

[2] The translation is done by the author as close as possible to original wording.

place of work and a narrower time frame. Mostly, students can decide with whom they work and how much they invest or participate. On a level 2 inquiry students can plan the investigation independently. They choose materials and methods. Teachers' scaffolding has to concentrate much more on achieving the subject-oriented learning goal than in the *Lernwerkstatt*, i.e. students are supposed to understand a certain chemical concept. This is especially evident in IRE and inquiry level 1 settings. From level 1 to level 2 settings scaffolding on processual issues additionally becomes important. Especially, scaffolding on planning and conducting an experiment is dominant in level 2 inquiries. This means, methodical aims are strived for next to subject-matter knowledge. It becomes evident, however, that the teacher's scaffolding is more amenable to systemic conditions which is less perceived in the *Lernwerkstatt*. For example, the students need to meet the subject-specific end in the determined time of one or two school lessons. The following transcript illustrates the restriction. The task of the students is to separate water and ethanol and how to contain the ethanol. They decide how to solve the problem with the support of the teacher.

Student 1: yes, but this is not good, because it drops down from up there
Teacher: ok and what can we do that it
Student 2: I just take a bigger jar with a bigger diameter than that one
Student 1: no, one could do it with this (points to a glass tube)
Teacher: how could one shorten this way? (points from glass tube to beaker)
Student 2: by adding another glass tube
Teacher: how else?
Student 2: by putting this higher
Student 3: no, you just have to put this higher
Student 2: look (puts a measuring cylinder on a beaker)
Student 3: yes exactly
Teacher: get a tin pot from over there
Student 3: tin pot, fast, tin pot
Student 1: no, I have a better idea, can one this here //
Student 3: // tin pot
Student 1: no, student 3, look we do it like this (see figure 3)
Teacher: Student 1, you get bogged down in details again, you have only 15 minutes left

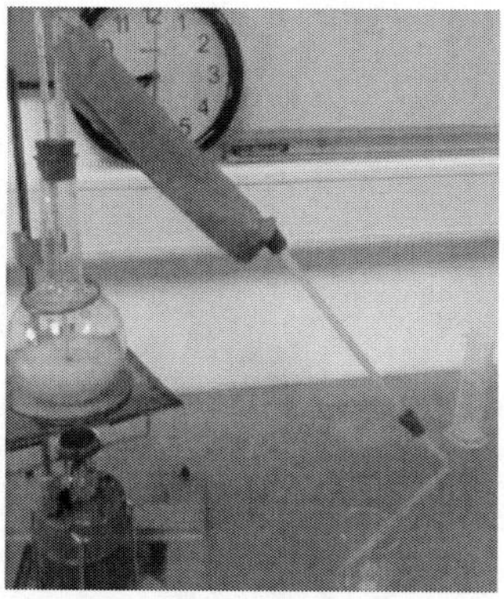

Figure 3. Distillation.

Following the terminology of Häcker (2007) the students work "self-determined" in the *Lernwerkstatt* whereas in chemistry they work "self-regulated". While self-regulation allows for procedural decisions, self-determination also allows for decisions about aims and the content students want to engage in. Self-determination demands of the teachers to trust the students in making good decisions and to be open in terms of learning goals which will be achieved. This classifies *Lernwerkstatt* as an inclusive approach (see above). In chemistry the curriculum sets the content-specific learning goals and by using settings like inquiry level 2 the teacher facilitates self-regulation.

This leads to a second major category: the sway of subject-matter on scaffolding which is described in the following part.

The Sway of Subject-Matter Knowledge on Scaffolding

One major difference between *Lernwerkstatt* and chemistry lessons is the role of subject matter knowledge. Although both formats are skill-led and students are supposed to acquire methodical, social and personal competencies, chemistry is much more content-driven. All tasks, independent of the teaching approach, strive for the understanding of the same chemical concept, whereas in *Lernwerkstatt* the teachers do explicitly not strive for achieving specific learning goals in the content area. That does not mean that students are not intended to learn content, but the learning gain in subject matter depends on the individual project and is not set beforehand. The difference is remarkable in that the chemistry teacher at this school is very aware of the fact that she does not teach all the content she is supposed to according to the chemistry curriculum. The following transcript is taken from an informal conversation between the teacher and the researcher.

> Teacher: I only teach a quarter of the content, but therefore it is, I think, more understandable for the kids and I just accept that. Because also when I teach all the content, they don't memorize it.

To help the students memorize and make the content more meaningful, the teacher draws on inquiry-based settings, namely level 1 and 2, which alternate with IRE sessions. In the IRE phases the practical work is prepared or evaluated subject-specifically and methodically. Nevertheless, students are supposed to learn the same content, naturally they learn in varying degrees or with additional support. For example, in the lesson about distillation (see figure 3) all students shall understand that substances can be separated according to their features, here the different boiling temperature. In the *Lernwerkstatt* all students learn something about the topics light and color, water or insects, for example, but they all learn different content which is appreciated in the *Lernwerkstatt.* According to Feuser (1995) a joint topic, but different achievements make learning inclusive. This classifies *Lernwerkstatt* as a more inclusive approach in the sense of Florian and Black-Hawkins (2011) than subject-specific lessons.

The subsequent question is if chemistry has reached its limit of inclusive potential with self-regulated inquiries or if the limit can be extended like in the *Lernwerkstatt.* To answer the question, one has to know that chemical investigations do not take place in the *Lernwerkstatt* when there is no input in this direction. For example, in the *Lernwerkstatt* with the topic "small life big time" (insects) students phrased the question "What is honey made of?". The investigations became chemical when a coach informed them about the possibility of detecting glucose and fructose with a Fehling's test and gave instructions how to proceed. As chemical procedures are often not part of everyday knowledge and chemistry starts in 8th grade (in Austria), students ought to have a lot of prior knowledge to work with chemical concepts in the *Lernwerkstatt*. Consequently, chemistry teachers need to provide information and structured settings in their lessons to establish prior knowledge. It would be interesting to implement *Lernwerkstatt* in upper secondary classes when students also have more chemical knowledge available. Students could learn to develop follow-up questions (coupled inquiry) which could lead to chemistry as a more inclusive approach in the sense of Florian and Black-Hawkins (2011). Consequently, either a curriculum change in this direction would be necessary as different learning goals ought to be allowed. Or another possibility that does not require a curriculum change would be to intensify the collaboration between subjects and formats like *Lernwerkstatt*. Subject-specific teaching could implement inquiry successively (level 0 to 2) and develop prior knowledge so that students can come with their open questions into the *Lernwerkstatt*. Of course, *Lernwerkstatt* should then not be offered just once a year.

Conclusion

The school cooperating in this research project addresses inclusion as holistic school development. However, this does not mean that lessons are totally and automatically inclusive. Teachers need to find inclusive pedagogical approaches and scaffold them carefully. This is challenging in subject-specific teaching in another way than in interdisciplinary project-oriented formats. Science and chemistry show to be subjects that put high requirements on being inclusive. Even at this school with a long tradition in inclusive teaching teachers still work on dealing with the demands. Their secret is the motivation for ongoing development inspired by reflection and collaboration.

The science teachers in this project decided for inquiry-based learning which is recommended as an inclusive approach. Data analysis showed that interdisciplinary, skill-led, project-oriented formats based on self-determined learning like *Lernwerkstatt*, i.e. open inquiry with carefully structured scaffolding, can be classified as inclusive pedagogies. In contrast, chemistry lessons are much more amenable to systemic conditions, e.g. a curriculum with content prescriptions or at least expectations, grading, time frames, safety regulations which lead to exclusion etc. Thus, despite the use of inquiry settings and an approximation to inclusive approaches, chemistry lessons do not explicitly allow for differentiated learning aims and full self-determination.

A concluding hypothesis is that the subject culture and the necessitation to teach content-driven is not compatible with full inclusion. Teachers can only try to approach it as much as possible.

Grounded Theory has shown to be a useful and appropriate methodology for science education research as it analyses social phenomena systematically and comparatively. Thus, it allows for deep insights into authentic inclusive teaching and for the identification of scaffolding patterns and restrictions. Subject-specific facilitating factors and barriers of inclusion can be explored in a detailed manner. A limit of the methodology could be seen in the low number of classes that can be contrasted. Further case studies in different schools, subjects and age levels enabled by the collaboration of researchers are necessary which have to be merged and contrasted. However, as the methodology seeks for generating hypotheses, it facilitates to abstract from the individual case and to develop valid theories for science education research.

ACKNOWLEDGMENTS

I am very thankful for the school and teachers cooperating intensively in this research project. Furthermore, I thank Sandra Puddu for her constructive comments on this chapter.

REFERENCES

Abels, S. (2014a). Inklusiver naturwissenschaftlicher Unterricht in der Lernwerkstatt Donaustadt. *Schulpädagogik heute, 5*(10).

Abels, S. (2014b). Inquiry-based Science Education and Special Needs – Teachers' Reflections on an Inclusive Setting. *Sisyphus - Journal of Education, 2*(2), 124-154.

Abels, S. (accepted). Implementing Inquiry-based Science Education to Foster Emotional Engagement of Special Needs Students. In M. Kahveci & M. Orgill (Eds.), *Affective Dimensions in Chemistry Education* (pp. 26). Heidelberg: Springer.

Abels, S. & Markic, S. (2013). Umgang mit Vielfalt – neue Perspektiven im Chemieunterricht. *Naturwissenschaften im Unterricht - Chemie, 24*(135), 2-6.

Abrams, E., Southerland, S. A. & Evans, C. (2008). Introduction. Inquiry in the classroom: Identifying Necessary Components of a Useful Definition. In E. Abrams, S. A. Southerland & P. Silva (Eds.), *Inquiry in the classroom. Realities and Opportunities* (pp. xi-xlii). Charlotte, North Carolina: Information Age Publishing.

Achieve Inc. on behalf of the twenty-six states and partners that collaborated on the NGSS. (2013). Next Generation Science Standards. Retrieved 18/07/2013, from http://www.nextgenscience.org/next-generation-science-standards

Ahlring, I. (2000). Es führen viele Wege nach Rom... Muster und Module binnendifferenzierenden Unterrichts. *Praxis Schule 5-10, 11*(6), 8-14.

Avramidis, E. & Norwich, B. (2002). Teachers' attitudes towards integration / inclusion: a review of the literature. *European Journal of Special Needs Education, 17*(2), 129-147.

Bakken, J. P., Mastropieri, M. A. & Scruggs, T. E. (1997). Reading Comprehension of Expository Science Material and Students with Learning Disabilities: a Comparison of Strategies. *The Journal of Special Education, 31*(3), 300-324.

Blanchard, M. R., Southerland, S. A., Osborne, J. W., Sampson, V. D., Annetta, L. A. & Granger, E. M. (2010). Is Inquiry Possible in Light of Accountability?: A Quantitative Comparison of the Relative Effectiveness of Guided Inquiry and Verification Laboratory Instruction. *Science Education, 94*(4), 577-616.

Breuer, F. (2010). *Reflexive Grounded Theory. Eine Einführung für die Forschungspraxis* (2 ed.). Wiesbaden: VS Verlag für Sozialwissenschaften.

Breuer, F. (2011). Predecessor-Successor Transitions in Institutional and Interpersonal Contexts. On the Development of a Theory of the Transfer of Personal Objects. *Forum Qualitative Sozialforschung / Forum: Qualitative Social Research, 12*(2). http://www.qualitative-research.net/index.php/fqs/article/view/1653/3269

Brigham, F. J., Scruggs, T. E. & Mastropieri, M. A. (2011). Science Education and Students with Learning Disabilities. *Learning Disabilities Research & Practice, 26*(4), 223-232.

Browder, D. M., Trela, K., Courtade, G. R., Jimenez, B. A., Knight, V. & Flowers, C. (2012). Teaching Mathematics and Science Standards to Students With Moderate and Severe Developmental Disabilities. *The Journal of Special Education, 46*(1), 26-35.

Charmaz, K. (2006). *Constructing Grounded Theory: A Practical Guide Through Qualitative Analysis*. London: Sage.

Dexter, D. D., Park, Y. J. & Hughes, C. A. (2011). A Meta-Analytic Review of Graphic Organizers and Science Instruction for Adolescents with Learning Disabilities: Implications for the Intermediate and Secondary Science Classroom. *Learning Disabilities Research & Practice, 26*(4), 204-213.

Ellis, E. S. (1993). Integrative Strategy Instruction: A Potential Model for Teaching Content Area Subjects to Adolescents with Learning Disabilities. *Journal of Learning Disabilities, 26*(6), 358-383.

European Agency for Development in Special Needs Education. (2012). *Raising Achievement for All Learners – Quality in Inclusive Education*. Odense, Denmark: European Agency for Development in Special Needs Education.

Feuser, G. (1995). *Behinderte Kinder und Jugendliche: zwischen Integration und Aussonderung*. Darmstadt: Wiss. Buchges.

Feyerer, E. (2007). Integration an (ober-)österreichischen Hauptschulen Eine Standortbestimmung für das Projekt Schulentwicklung durch Schulprofilierung. *Zeitschrift für Inklusion, 1*, 1-21.

Florian, L. & Black-Hawkins, K. (2011). Exploring inclusive pedagogy. *British Educational Research Journal, 37*(5), 813-828.

Furtak, E. M. (2008). *The Dilemma of Guidance. An Exploration of Scientific Inquiry Teaching*. Saarbrücken: VDM Verlag Dr. Müller.

Gilbert, J. K. (2006). On the Nature of "Context" in Chemical Education. *International Journal of Science Education, 28*(9), 957-976.

Goldman, R. (2007). Video representations and the perspectivity framework: epistemology, ethnography, evaluation and ethics. In R. Goldman, R. Pea, B. Barron & S. J. Denny (Eds.), *Video Research in the Learning Sciences* (pp. 3-37). Mahwah, New Jersey, London: Lawrence Erlbaum Associates.

Griffiths, C. (2013). Using Grounded Theory to Analyze Qualitative Observational Data that is Obtained by Video Recording. *Grounded Theory Review. An International Journal, 12*(1).

Häcker, T. (2007). Portfolio - ein Medium im Spannungsfeld zwischen Optimierung und Humanisierung des Lernens. In M. Gläser-Zikuda & T. Hascher (Eds.), *Lernprozesse dokumentieren, reflektieren und beurteilen. Lerntagebuch und Portfolio in Forschung und Praxis* (pp. 63-85). Bad Heilbrunn: Klinkhardt.

Hagstedt, H. (2004). "Fordernde" Lernorte: Lernwerkstätten. *Die Grundschulzeitschrift, 18*(171), 48-50.

Hart, S., Drummond, M. & McIntyre, D. (2007). Learning without limits. In L. Florian (Ed.), *The Sage handbook of special education* (pp. 499-514). London: Sage.

Higgins, S., Kokotsaki, D. & Coe, R. (2011). Toolkit of Strategies to Improve Learning. Summary for Schools Spending the Pupil Premium. http://www.cem.org/attachments/1toolkit-summary-final-r-2-.pdf

Hofstein, A., Navon, O., Kipnis, M. & Mamlok-Naaman, R. (2005). Developing Students' Ability to Ask More and Better Questions Resulting from Inquiry-Type Chemistry Laboratories. *Journal of Research in Science Teaching, 00*(00), 1-16.

Janík, T., Seidel, T. & Najvar, P. (2009). Introduction: On the Power of Video Studies in Investigating Teaching and Learning. In T. Janík & T. Seidel (Eds.), *The Power of Video Studies in Investigating Teaching and Learning in the Classroom* (pp. 7-19). Münster u.a.: Waxmann.

Jimenez, B. A., Browder, D. M., Spooner, F. & Dibiase, W. (2012). Inclusive Inquiry Science Using Peer-Mediated Embedded Instruction for Students with Moderate Intellectual Disability. *Exceptional Children, 78*(3), 301-317.

Krell, G., Riedmüller, B., Sieben, B. & Vinz, D. (2007). Einleitung - Diversity Studies als integrierende Forschungsrichtung. In G. Krell, B. Riedmüller, B. Sieben & D. Vinz (Eds.), *Diversity Studies. Grundlagen und disziplinäre Ansätze* (pp. 7-16). Frankfurt a. M., New York: Campus.

Lederman, N. G., Southerland, S. A. & Akerson, V. L. (2008). Students' Knowledge and Skill with Inquiry. In E. Abrams, S. A. Southerland & P. Silva (Eds.), *Inquiry in the Classroom. Realities and Opportunities* (pp. 3-38). Charlotte, North Carolina: Information Age Publishing.

Markic, S. & Abels, S. (2013). Die Fachsprache der Chemie. Ein gemeinsames Anliegen von heterogenen Klassen. *Naturwissenschaften im Unterricht - Chemie, 24*(135), 10-14.

Markic, S. & Abels, S. (2014). Heterogeneity and Diversity – A Growing Challenge or Enrichment for Science Education in German Schools? *EURASIA Journal of Mathematics, Science & Technology Education, 10*(3), 219-227.

Martin-Hansen, L. (2002). Definig inquiry. Exploring the many types of inquiry in the science classroom. *The Science Teacher, 69* (2), 34-37.

Mason, L. H. & Hedin, L. R. (2011). Reading Science Text: Challenges for Students with Learning Disabilities and Considerations for Teachers. *Learning Disabilities Research & Practice, 26*(4), 214-222.

Mastropieri, M. A., Scruggs, T. E. & Butcher, K. (1997). How Effective is Inquiry Learning for Students with Mild Disabilities? *The Journal of Special Education, 31*, 199-211

Meijer, C. J. W. (2010). Special Needs Education in Europe: Inclusive Policies and Practices. *Zeitschrift für Inklusion, 4*(2). http://www.inklusion-online.net/index.php/inklusion-online/article/view/136/136

Minnerop-Haeler, L. (2013). Die Lernwerkstatt Donaustadt. Ein Beispiel für gelebte Inklusion. *Naturwissenschaften im Unterricht - Chemie, 24*(135), 36-39.

National Research Council. (2000). *Inquiry and the National Science Education Standards.* Washington D.C.: National Academy Press.

Nilsson, L. (2012). Using video methods in grounded theory research. In V. B. Martin & A. Gynnild (Eds.), *Grounded theory: the philosophy, method and work of Barney Glaser* (pp. 103-116). Boca Raton, Florida: Brown Walker Press.

Puddu, S., Keller, E. & Lembens, A. (2012). Potentials of Lernwerkstatt (Open Inquiry) for Pre-service Teachers' Professional Development. E-Book Proceedings of the ESERA 2011 Conference: Science learning and Citizenship. Part 12 (co-ed. D. Psillos & R. M. Sperandeo). In C. Bruguière, A. Tiberghien & P. Clément (Eds.), *ESERA 2011 Conference* (pp. 153-159). Lyon, France: European Science Education Research Association.

Rocard, M., Csermely, P., Jorde, D., Lenzen, D., Walberg-Henriksson, H. & Hemmo, V. (2007). Science Education Now: A Renewed Pedagogy for the Future of Europe. Retrieved 18 Mar 2008, from http://ec.europa.eu/research/science-society/document_library/pdf_06/report-rocard-on-science-education_en.pdf

Roth, K. J. (2009). Using Video Studies to Compare and Understand Science Teaching: Results from the TIMSS Video Study of 8th Grade Science Teaching. In T. Janík & T. Seidel (Eds.), *The Power of Video Studies in Investigating Teaching and Learning in the Classroom* (pp. 23-38). Münster, New York, München, Berlin: Waxmann.

Sander, W. (2008). Von der Einzelstunde zur längerfristigen Unterrichtsanalyse - Unterrichtsforschung nach der Grounded Theory. In GPJE (Ed.), *Diversity Studies und politische Bildung. Schriftenreihe der Gesellschaft für Politikdidaktik und Politische Jugend- und Erwachsenenbildung; 7* (pp. 113-123). Schwalbach/Ts.: Wochenschau.

Scruggs, T. E. & Mastropieri, M. A. (2000). The Effectiveness of Mnemonic Instruction for Students with Learning and Behavior Problems: An Update and Research Synthesis. *Journal of Behavioral Education, 10*(2-3), 163-173.

Scruggs, T. E. & Mastropieri, M. A. (2007). Science Learning in Special Education: The Case for Constructed Versus Instructed Learning. *Exceptionality, 15*(2), 57-74.

Scruggs, T. E., Mastropieri, M. A. & Okolo, C. M. (2008). Science and Social Studies for Students With Disabilities. *Focus on Exeptional Children, 41*(2), 1-24.

Seidel, T., Prenzel, M., Rimmele, R., Schwindt, K., Kobarg, M., Herweg, C., . . . Allolio-Näcke, L. (2006). Unterrichtsmuster und ihre Wirkungen. Eine Videostudie im Physikunterricht *Untersuchungen zur Bildungsqualität von Schule. Abschlussbericht des DFG-Schwerpunktprogramms* (pp. 99-123). Münster: Waxmann.

Sjøberg, S. & Schreiner, C. (2010). The ROSE project. An overview and key findings. Retrieved 11 Feb 2014, from http://roseproject.no/network/countries/norway/eng/nor-Sjoberg-Schreiner-overview-2010.pdf

Sliwka, A. (2010). From homogeneity to diversity in German education. In OECD (Ed.), *Educating Teachers for Diversity: Meeting the Challenge* (pp. 205-217): OECD Publishing.

Spooner, F., Knight, V. F., Browder, D. M. & Smith, B. R. (2012). Evidence-Based Practice for Teaching Academics to Students With Severe Developmental Disabilities. *Remedial and Special Education, 33*(6), 374-387.

Stroh, M. (2014). Inklusion im naturwissenschaftlichen Unterricht – Beschreibung eines Spannungsfeldes. *Schulpädagogik heute, 5*(10), 1-12.

Stuckey, M., Hofstein, A., Mamlok-Naaman, R. & Eilks, I. (2013). The meaning of 'relevance' in science education and its implications for the science curriculum. *Studies in Science Education, 49*(1), 1-34.

Sullivan Palincsar, A., Magnusson, S. J., Collins, K. M. & Cutter, J. (2001). Making Science Accessible to All: Results of a Design Experiment in Inclusive Classrooms. *Learning Disability Quarterly, 24*(1), 15-32.

Therrien, W. J., Taylor, J. C., Hosp, J. L., Kaldenberg, E. R. & Gorsh, J. (2011). Science Instruction for Students with Learning Disabilities: A Meta-Analysis. *Learning Disabilities Research & Practice, 26*(4), 188-203.

Tobin, R. & Tippett, C. (2013). Possibilities and Potential Barriers: Learning to Plan for Differentiated Instruction in Elementary Science. *International Journal of Science and Mathematics Education*, 1-21.

UNESCO. (2005). Guidelines for Inclusion: Ensuring Access to Education for All. Retrieved 11 Jan 2012, from http://unesdoc.unesco.org/images/0014/001402/140224e.pdf

United Nations. (2006). Convention on the Rights of Persons with Disabilities. Retrieved 21 Feb 2012, from http://www.un.org/disabilities/documents/convention/convoptprot-e.pdf

van der Valk, T. & de Jong, O. (2009). Scaffolding Science Teachers in Open-inquiry Teaching. *International Journal of Science Education, 31*(6), 829-850.

van Zee, E. & Minstrell, J. (1997). Using Questioning to Guide Student Thinking. *Journal of the Learning Sciences, 6*(2), 227-269.

Vygotsky, L. S. (1978). *Mind in Society: The Development of Higher Psychological Processes*. MA: Harvard University Press.

Weber, L. (1977). An Approach to Natural Materials. In R. Dropkin (Ed.), *The Teacher as Learner* (pp. 28-32). New York: City College Workshop Center.

Werning, R. & Lütje-Klose, B. (2007). Entdeckendes Lernen. In U. Heimlich & F. Wember (Eds.), *Didaktik des Unterrichts im Förderschwerpunkt Lernen. Ein Handbuch für Studium und Praxis* (pp. 149-162). Stuttgart: Kohlhammer.

Wildt, J. (2003). Reflexives Lernen in der Lehrerbildung. Ein Mehrebenenmodell in hochschuldidaktischer Perspektive. In A. Obolenski & H. Meyer (Eds.), *Forschendes Lernen. Theorie und Praxis einer professionellen LehrerInnenausbildung* (pp. 71-84). Bad Heilbrunn / Obb.: Klinkhardt.

Zocher, U. (2000). *Entdeckendes Lernen lernen. Zur praktischen Umsetzung eines pädagogischen Konzepts in Unterricht und Lehrerfortbildung*. Donauwörth: Auer

Reviewed by Mag. Sandra Puddu, University of Vienna

In: New Developments in Science Education Research
Editor: Nathan L. Yates
ISBN: 978-1-63463-914-9

Chapter 6

FREDERICK ACCUM (1769-1838) AND THE APPLICATION OF CHEMISTRY TO SOCIAL PROBLEMS

W. P. Palmer
Curtin University of Technology, Perth, WA, Australia

The purpose of this study is to celebrate the life and work of Frederick Accum as an exemplar of the importance of chemistry in the lives of ordinary people. Frederick Accum was born on March 29th, 1769 in Bückeburg, Germany. His father was a soap boiler and merchant and this may explain his early interest in chemistry. He became an apprentice to the firm of Brande, who had businesses in both Germany and England. In 1793, he left Germany to work as assistant chemist in Brande's London Pharmacy. After resigning from Brande's, he set up in business on his own as a chemical analyst. On 10th May, 1798, he married Mary Ann Simpson, with whom he had eight children and of these only two survived to adulthood.

In 1801 he was proposed by Count Rumford as Assistant Chemical Operator at the Royal Institution. A portrait of Accum as a young man shows him to be very finely dressed in the latest style and the fact that he established himself with influential friends demonstrated considerable social skill. In 1803, Accum published his first book, *System of Theoretical and Practical Chemistry* which was based on Lavoisier's system of nomenclature. Over his lifetime he wrote about a dozen books and many chemical papers.

Also in 1803, Accum resigned from his position at the Royal Institution and instead lectured at the Surrey Institution. At the same time he gave private lectures from his own home and teamed up with another German émigré, Friedrich Winsor, who had a plan for lighting London streets and buildings using gas lighting. Accum was also fascinated by the possibilities and had already experimented on this new technology. He gave evidence on gas lighting to Committees of the House of Commons and of the House of Lords as a chemical expert.

His business prospered; he assisted a number of Americans such as Benjamin Silliman through his lectures and by selling them equipment to start university laboratories. In 1812 he became a director of the newly formed Gas Light and Coke Company. However by 1814 there was a severe economic downturn in Britain which harmed his business. In 1816, his wife, Mary died.

Accum, as an analyst, became aware of the problem harmful substances being added to food by manufacturers and wrote a book entitled A Treatise on Adulterations of Food and Culinary Poisons. This received great public acclaim but brought him many enemies as he named companies selling dangerous products. Shortly after the publication of this book, he was accused of damaging books at the Royal Institution Library. The final result was that he was so depressed that he hastily returned to Germany, where he worked as an academic for the remainder of his life; Accum died on 28th June 1838.

SOURCES

Friedrich Christian Accum has been the subject of numerous articles over the past century, and he is the subject of considerable interest as each new generation sees the important issues on which he concentrated his research effort in a different light. Most one volume dictionaries of science, such as Williams (1994), do not mention him. On the other hand, his life and work has been well-documented by historians of science, particularly Charles A. Browne (Deischer, 1948) who worked on Accum's biography for many years. Accum's importance in this International Year of Chemistry (2011) is that he was one of the first chemists to apply his chemical knowledge to the practical problems of his day. More than that, he was amongst the first in the United Kingdom to earn his living through chemistry, rather than through title or inherited wealth.

The major primary sources were Accum's own writings of more than fifteen books which include Accum (1798; 1803a; 1803b; 1814; 1815; 1819; 1820; 1824). The World Catalogue (http://www.worldcat.org/) claims that Accum published '267 works in 571 publications in 8 languages with 4,053 library holdings'. It can also be seen that many of his publications or new editions of old works were published after 1820, when he left England. Accum's work was of a high quality and his *Practical Treatise on Gas Light*, which sold for twelve shillings (TLM, 1815), has been described as 'one of the most elegant textbooks in the English language' (Klingender, 1975, p. 149). His last book (Accum, 1824), published anonymously, is described as a 'remarkable and profusely illustrated treasure house of information' (Brock, 2000, p. 188). The books he wrote would have supported his business financially.

A wonderful feature of the internet is that many chemistry textbooks are available to view online through organisations such as The Hathi Trust Digital Library (http://www.hathitrust.org/), the Open Library (http://openlibrary.org/books) or the Internet Archive (American Libraries) (http://www.archive.org/). Also many of Accum's works are available moderately cheaply as instant print volumes. Major secondary sources for biographical information were Bing (1966), Browne (1925 & 1948), Cole (1951), Fullmer, (1967), Gee (2004), Sumner (2007) and Thackray (1970). The Wikipedia entry is a good online source of information about Accum, (URL: Friedrich_Accum). Other works utilised were articles and advertisements found in newspapers and journals of the early nineteenth century, together with more recent commentary.

HIS EARLY LIFE

Frederick Accum was born on March 29th, 1769 in Bückeburg, Germany, about fifty kilometres from Hanover. His father was born in the Jewish faith with the name of Herz Marcus (Cole, 1951. p. 128) but he changed faith in order to marry Judith La Motte, who was a Huguenot; they were married in 1755. The name 'Accum' chosen by Friedrich's father indicated his change of faith as its meaning is 'Gentile' or 'non-Jew' in Hebrew (Browne, 1948). Friedrich Christian was the sixth child of seven of whom only three survived to adulthood (Thackray, 1970, p. 43); his father died on 9th May 1772, when he was just three years old (Browne, 1925, p. 831)

Accum's father was a soap boiler and merchant and this may explain his early interest in chemistry as the family continued the business after his father's death. The income from the business was enough for him to receive a good education at the local Gymnasium (Browne, 1925, p. 832). His passion for hard work and strong moral principles were the result of his mother's care. He became an apprentice to the firm of Brande, who had businesses in both Germany and England and who were apothecaries to King George III of England. In 1793, he left Germany to work as assistant chemist in Brande's London Pharmacy. He undertook further studies in science and medicine, becoming acquainted with Anthony Carlisle (surgeon) and William Nicholson (chemist and editor of *Nicholson's Journal*) (Bing, 1966, p. 4); these two men remained his most loyal friends and supporters. Accum's friendship with Nicholson led to him writing a number of papers for *Nicholson's Journal* (Browne, 1925, p. 848); several of these related to his pharmaceutical knowledge, for example on the genuineness and purity of drugs (Accum, 1798). Accum presented Nicholson with a small sample of beet sugar from Germany: 'Mr. Accum has presented me with samples of this sugar received from Berlin, where I understand it is now very commonly manufactured' (Nicholson, 1800); sugar from beet later became an industry of huge economic importance (Browne, 1925, p. 835). Accum's early interest in the safety of food is thus evident early in his career. Much later in 1820, Accum was called as an expert witness with many of the accomplished chemists of the time (Brande, Allen, Parkes, Garden and Faraday) in a case about a fire at the premises of a sugar manufacturer (BNP, 1820; Kargon, 1986, pp. 20-21).

A few years later, Accum resigned from Brande's and set up in business on his own as a chemical analyst. On 10th May, 1798, he married Mary Ann Simpson, with whom he had eight children and of these only two survived to adulthood. In 1800 he moved from 17 Haymarket to 11 Old Compton Street, Soho, where he started his business of selling chemical apparatus and established a laboratory; he operated his business from here for the next twenty years, some of them in partnership with Alexander Garden. In 1801 he was proposed by Count Rumford as Assistant Chemical Operator at the Royal Institution as assistant to Humphry Davy. The closeness of the London chemical community of the early nineteenth century may be illustrated by the fact that Accum, while assisting Davy, Professor of Chemistry at the Royal Institution also taught William Thomas Brande at St George's Hospital (James, 2004). When Davy resigned from the Royal Institution, William Thomas Brande was appointed the next Professor of Chemistry, often cooperating with Accum in professional work. Accum published his first book *A System of Theoretical and Practical Chemistry* in 1803. It was described as 'the first text book on general chemistry written in the English language to be based on Lavoisier's new principles' (Cole, 1951. p. 130) with 'some aspects of a trade catalogue' (Knight, 2003).

A portrait of Accum as a young man shows him to be very finely dressed in the latest style and the fact that he established himself with influential friends so quickly shows that he was a very personable young man.

His Later Career

In 1803, Accum resigned from his position at the Royal Institution, perhaps because his original proposer to the position, Count Rumford, resigned. Accum expanded his business and was appointed as a lecturer at the Surrey Institution. A course on chemistry which was the most systematic course on offer, was an important component of the Surrey Institution's courses and it was delivered by Frederick Accum during seven seasons (1808-1811 and 1818-1820) (Kurzer, 2000, p. 124). His business prospered; he assisted a number of Americans such as Benjamin Silliman, William Peck, and James Dana (Cole, 1951, p. 135) through his lectures and by selling them equipment to start university laboratories in America. Accum had support from some very influential people such as the Marquis of Landsdown, Sir Joseph Banks, Sir William Hamilton, Viscount Palmerston and Admiral Nelson (Browne, 1925, p. 1018) in purchasing his first chemistry book (Accum, 1803a). It is noticeable that his support does not come from practicing chemists but rather from a general public, anxious to learn more about chemistry.

Accum teamed up with another German émigré, Frederick Winsor, who had a plan for lighting London streets and buildings using gas lighting. How essential was Accum in establishing a viable company to provide gas lighting in London? There are three names associated with this project: these were Frederick Winsor, Frederic Accum and Samuel Clegg. Winsor (formerly Winzer) obtained the patent for the process in May 1804 (Browne, 1925, p. 1009). He then publicised his scheme to finance a public company to provide gas lighting in London which required parliamentary approval. Winsor, an 'ostentatious entrepreneur' (DiLaura, 2000, p. 183), was evidently a poor lecturer and wrote wildly exaggerated pamphlets to promote his scheme (Clow and Clow, 1992, p. 434). Accum had already carried out a series of experiments on this new technology with his partner, Alexander Garden (Ihde, 1966, pp. 452-453) and gave evidence to Committees of the House of Commons as a chemical expert in favour of Winsor's scheme. He showed the Committee a finely worked box which contained actual samples of the chemicals used and produced in making the gas used in illumination, which illustrates his technical and artistic abilities. However some of his testimony was ridiculed by opponents of the scheme and he appeared a mercenary foreigner (Browne, 1925, p. 1012) due to his requests for payment. The Parliamentary charter for such a company, promoted by Winsor, was defeated in 1809 (Goldfarb, 1971, p. 474). On resubmitting a smaller and more practical scheme, the Parliamentary charter was at last successful. In 1812, Accum became a director of the newly formed Gas Light and Coke Company with duties as a practical chemist. However neither Accum nor Winsor (Goldfarb, 1971, p. 474) had sufficient experience in engineering or administration and Samuel Clegg, an experienced practical engineer, took charge of day to day problems in 1813. This was a good appointment and by 1815 there were twenty six miles of gas main (Clow et al, 1992, p. 435). Winsor, Accum and eventually Clegg left the company within five years.

However by 1814 there was a severe economic downturn in Britain which harmed Accum's business. Nonetheless his research continued with work on the newly discovered element, iodine (Accum, 1814; TMP, 1814). In 1815 he showed that the low boiling point fractions of coal-tar were excellent paint thinners (Forbes and Dijksterhuis, 1963, p. 492). In 1816, his wife, Mary died. Accum (Browne, 1948, p. 5) wrote of his intense sorrow that 'death has torn away my wife' after three weeks illness. Accum was left alone with his seventeen year old daughter; his son lived in Paris and Accum did not think his son would be interested in a chemical career. All these pressures took much of the pleasure from his life.

In the course of his business as an operational chemist and expert witness, Accum frequently gave evidence in court. In one case, the smell of chemicals was offensive to a resident; Accum did not think there was a public danger and his evidence favoured the defendant who was a manufacturer (TT, 1818). In 1819 he analysed the waters at Thetford Spa, writing a book about them (Accum, 1819). Accum's health was drunk and he was honoured by His Grace the Duke of Grafton who said that ‘Their [the waters] chemical analysis is a tower of strength in the name of Accum’ (BNP, 1819). His books were often advertised in newspapers, for example an advertisement for *Accum's chymical amusement* (TT, 1819). As an analyst, he became aware of the problem of harmful substances being added to food by manufacturers, where members of the public suffered injury through ingesting poisonous substances. He then published *A treatise on the art of brewing* (TT, 1820), which indicated malpractice by the brewers.

However the book, entitled *A Treatise on Adulterations of Food and Culinary Poisons* (Accum, 1820, pp. 15-16) in which he described the adulteration of many food staples including cream, confectionary, pepper, tea, coffee, spirituous liquors, milk, meat, vegetables and bread, was his major contribution. He found most to be intentionally adulterated, often with dangerous chemicals (Freckelton, 2009). His book was published in January 1820 and widely advertised, as for example in *The Morning Chronicle* (TMC, 1820). Many other papers contained reviews or summaries over the next few weeks with the news gradually spreading out from London, as for example in an article in the *Royal Cornwall Gazette* (RCG, 1820) and a cynical review reprinted from *Blackwood's Magazine* in the *Glasgow Herald* (GH, 1820).

Accum's chemical knowledge allowed him to analyse the impurities and was often able to explain how people could test foods for themselves. His style shows a good command of English and indicates his passion in fighting the wrongs done to ordinary people through dishonest merchants.

> It is really astonishing that the penal law is not more effectually enforced against practices so inimical to the public welfare. The man, who robs a fellow subject of a few shillings on the high-way, is sentenced to death while he who distributes a slow poison to a whole community, escapes unpunished (Accum, 1820, pp. 15-16).

Accum returned to Germany during the early part of 1821; the story of his final three months in Britain can be followed through newspaper reports, which indicate the intense pressure that he had to endure through adverse public opinion. Accum's book on food adulteration received great public acclaim but brought him many enemies as he named the companies that sold dangerous products. Shortly after the publication of this book, he was accused of damaging books at the Royal Institution Library; he was evidently observed ripping the pages out of some books by the assistant librarian, who had used a spy-hole in the paneling on 20th December, 1820. Accum's house was searched on the next day (Browne, 1925, p. 1141) and some loose pages torn from books were found. He was accused of theft but the magistrate found him ‘not guilty’ and dismissed the charge because the cost of the paper he was alleged to have stolen was of so little value (CM, 1820: TMP, 1820a). The Royal Institution decided to appeal against this verdict.

Two verses from the six verse poem may indicate the very cynical attitude of the press, when commentating on these events.

The lecturer Fred
Took it into his head
To make war on butchers and cooks
But he happened to pass
Over one knavish class
The rogues who deteriorate books.

But Fred coolly replied
"You ought not to chide
Tho' Directors may give themselves airs,
You who join to defame
My works, cannot blame
Me, for taking a leaf out of theirs.
(TMP, 1820b)

Accum was bound over and granted bail when he appeared at the Westminster Sessions on January 4, 1821 (TT, 1821a). A letter pleading for leniency for Accum appeared, signed A.C (probably Anthony Carlisle) (TT, 1821b), but it had no effect. On 9th January 1821, Accum wrote to the *Morning Post*, proclaiming his innocence and asking the public to suspend their judgement of him (Accum, 1821). He failed to appear at the Westminster court on 6 April (TT, 1821c), having left for Germany secretly prior to the court case. On 16th April he put many of his possessions up for auction (TT, 1821d). So many things had gone badly for him in the previous few years that had depressed him that he must have felt that a new start in Germany was the only solution. He worked as an academic in Germany for the remainder of his life, where he held two different professorships simultaneously and also found time to write a book on the physical and chemical properties of building materials. He also appeared to have retained a large portion of his wealth, transferred from London (Browne, 1948, p. 7).

His earlier books were reprinted and published anonymously or with Mucca, as a nom-de-plume. He died in Berlin (Anon, 1838) on 28th June 1838 aged 69 years (Cole, 1951 p. 142).

REVIEWS AND THE OPINION OF CONTEMPORARIES

Accum had been Davy's assistant at the Royal Institution. In Humphry Davy's opinion, Accum was a 'cheat and a quack' (Hamlin, 1990) quoting Fullmer (1967, p. 146) whose opinion was that it only referred to Davy's opinion of Accum's support of Winsor's plans for financing his gas-lighting scheme for London. In his experimental work, Accum was distinctly sloppy, having little regard for his own safety (Accum, 1803b, p.2) or that of his pupil, Benjamin Silliman (quoted Collins, 2007, p. 60). The frontispiece in *One Thousand Experiments in Chemistry* (McKenzie, 1820) shows the conditions that the workers endured (Knight, 2003) in a hellish gasworks, so part of Accum's legacy was also of promoting dangerous industrial conditions. Accum was very careful with money and some instances of apparent meanness can be quoted. Silliman tells the story that when Accum had a contract with the Prime Minister, William Pitt, to supply chemical apparatus to India, he then supplied all the old apparatus in his store (Browne, 1925, pp. 846-847: Fulton & Thomson, 1947, p. 57) and thought this quite funny. On the first occasion where he gave evidence to a House of Commons Committee, he indicated he was not prepared to divulge further technical information without being paid. As a postscript to his second article about iodine which was only to acknowledge a mistake in the earlier article he asked readers to be sure to pay the postage before writing to him (Accum, 1914, p. 211). On the other hand, Silliman and Peck said how generous he had been with them in giving them tuition time (Browne, 1925, p. 841). Possibly his worst academic fault was plagiarism; both Sumner (2007, pp. 35-37) and Wilson (2008, pp. 43-44) show that some sections of his books were copied from older works without attribution.

Conclusion

Accum's story is of interest in itself, but many aspects illustrate the interplay between science, culture and society. His work in introducing gas lighting to London was important. His work in preventing the adulteration of food was ground-breaking, but not followed through, so had little practical effect. He was an early chemist who applies his chemical knowledge to construct an identity for himself. The value of friendship in his life is apparent but the enemies that he made managed to destroy his reputation and his business, but he survived and died a rich and successful man. In this International Year of Chemistry, chemists should celebrate the life of a man who worked with scientists from many nations.

References

Accum, F. (1798). An attempt to discover the genuineness and purity of drugs and medical preparations. *Journal of Natural Philosophy, Chemistry, and the Arts* (Nicholson's Journal). 2, pp. 118- 122.

Accum, F. (1803a). *A System of Theoretical and Practical Chemistry*. London: from the Press of G. Hayden.

Accum, F. (1803b). Experiments and Observations on the Compound of Phosphorus and Sulphur and the Dangerous Explosions It Makes When Exposed to Heat, *A journal of natural philosophy, chemistry and the arts* (Edited, Nicholson, W.) September, pp.1-7.

Accum, F. (1814). Further Observations concerning the Production of the singular Substance called lode, or Iodine, *Tilloch's Philosophical Journal*, Vol 43, pp. 141- 146 and pp. 210-211.

Accum, F. (1815). *A practical treatise on gas-light: Exhibiting a summary description of the apparatus and machinery*, London: R. Ackermann.

Accum, F. (1819). *Guide to the chalybeate spring of Thetford: Exhibiting the general and primary effects of the Thetford spa*, London: Sold by T. Boys.

Accum, F. (1820). *A treatise on adulterations of food, and culinary poisons: Exhibiting the fraudulent sophistications of bread, beer, wine, spiritous liquors, tea, coffee, cream, confectionery, vinegar, mustard, pepper, cheese, olive oil, pickles, and other articles employed in domestic economy, and methods of detecting them*. London: Longman.

Accum, F. C. (1824). *An explanatory dictionary of the apparatus and instruments employed in the various operations of philosophical and experimental chemistry*. London: Printed for Thomas Boys.

Anon (1838). Mr. Frederick Accum, *The Gentleman's magazine,* Volume 165, October, p. 448.

Bing, F. C. (1966). Fredrick Accum, a biographical sketch, *The Journal of Nutrition*, 89(1) at URL: http://jn.nutrition.org/content/89/1/2.full.pdf

Brock, W. H. (2000). *The chemical tree: a history of chemistry*. New York: W. W. Norton & Company Ltd.

Browne, C. A. (1925). The life and chemical services of Fredrick Accum, *Journal of Chemical Education 2* (10), 829-851, *2* (11), 1008- 10034 and *2* (12), 1140-1149.

Browne, C. A. (1948). Recently acquired information concerning Fredrick Accum (1769-1838), *Chymia*, Volume 1, pp. 1-9.

Clow, A. & Clow, N. L. (1992). *The chemical revolution*. London: Batchworth Press Ltd (reprint from Gordon and Breach, Science Publishers, Reading).

Cole, R. J. (1951). Friedrich Accum: a biographical study, *Annals of Science, No 7*, p. 128.

Collins, P. (2007). Fake food and poison pills. *New Scientist;* (10/20/2007) 196(2626)60-61.

Deischer, C. K. (1948). A memorial tribute to Dr C. A. Browne, *Chymia*, Volume 1, pp. 11-22.

DiLaura, D. L. (2000). Farewell to the incandescent lamp: a retrospective to recount, assess, and honour a passing technology, *Leukos*, 5(3)183-235 (January)

Forbes, R. J. & Dijksterhuis, E. J., (1963). *A history of science and technology, Volume Two*, Harmondsworth, Penguin.

Friedrich_Accum at URL: http://en.wikipedia.org/wiki/Friedrich_Accum (Accessed on 12/02/2011).

Freckelton, I. (2009). Food law: challenges and future directions, *Deakin Law Review*, volume 14, No. 2.

Fullmer, J. Z. (1967). Davy's sketches of his contemporaries, *Chymia,* Volume 12, pp. 127-150.

Fulton, J. & Thomson, E. H. (1947). *Benjamin Silliman 1779-1864: Pathfinder in American Science*. New York: Henry Schuman.

Gee, B. (2004). Accum, Friedrich Christian (1769–1838), *Oxford Dictionary of National Biography*. Oxford: Oxford University Press.

Goldfarb, S. J. (1971).A Regency Gas Burner, Technology and Culture,12(3)471-476 (July)

Hamlin, C. (1990). *A Science of Impurity: Water Analysis in Nineteenth Century Britain*. Berkeley: University of California Press, pp. 52-55.

Ihde, A. (1966). *The development of modern chemistry*. New York: Harper & Row, pp. 438-454.

James, F. A. J. L. (2004). Brande, William Thomas (1788–1866), *Oxford Dictionary of National Biography*. Oxford: Oxford University Press.

Kargon, R. (1986). Expert testimony in historical perspective, *Law and Human Behavior*, 10(1-2)15-27.

Klingender, F. D. (1975). Art and the Industrial Revolution (edited and revised by Arthur Elton). Frogmore, St. Albans: Paladin.

Knight, D. (2003). Exalting understanding without depressing imagination': depicting chemical process, *HYLE--International Journal for Philosophy of Chemistry*, Volume 9, No.2, pp. 171-189.

Kurzer, F. (2000). A History of the Surrey Institution, *Annals of Science*, 57, 109- 141.

Nicholson, W.(1800). Sugar from white beet. *Journal of Natural Philosophy, Chemistry, and the Arts (Nicholson's Journal)* January, pp. 474-475.

Sumner, J. (2007). Retailing scandal: the disappearance of Friedrich Accum, in Caleb, A. M. (Editor) *(Re)creating science in nineteenth-century Britain.* Newcastle: Cambridge Scholars, Publishers, pp. 32-48.

Thackray, A. (1970). Accum, Friedrich Christian in Gillispie, C. C. (Editor) *Dictionary of scientific biography- Abailard- Berg Volume 1.* New York: American Council of Learned Societies and Charles Scribner's Sons, Publishers, pp. 43-44.

Williams, T. I. (1994). *Biographical dictionary of scientists.* Glasgow: Harper Collins Publishers.

Wilson, B. (2008). *Swindled: The Dark History of Food Fraud, from Poisoned Candy to Counterfeit Coffee.* Princeton University Press, pp. 1- 20.

REFERENCES FROM NEWSPAPERS

Accum, Frederick (1821). Letter to the editor, *TMP*, Tuesday, January 09, Issue 15543.

BNP (1819). Thetford Spa, BNP, Wednesday, September 22, Issue, 1943.

BNP (1820). Untitled- Adulteration of food and wine. BNP, Wednesday, January 26, Issue 1961.

BNP (1820). Postscript ,*BNP,* Wednesday, April 19; Issue 1973.

CM (1820). Police, *CM*, Thursday, December 28; Issue 15495.

GH (1820). From Blackwood's Magazine, *GH,* Monday, February 28, Issue 1793.

RCG (1820). Summary of A Treatise on Adulterations of Food and Culinary Poisons, *RCG,* Friday, January 29, Issue 866.

TLM (1815). Advertisement. *TLM*, Saturday, June 24; Issue 2609.

TMC (1820). Advertisement for *A Treatise on Adulterations of Food and Culinary Poisons,* TMC, Monday, January 10, 1820; Issue 15818.

TMP (1814). Remarkable discovery in chemistry, *TMP*, Tuesday, February 08; Issue 13432.

TMP (1820a). Police intelligence: Robbery at the Royal Institution. *TMP*, Friday, December 22; Issue 15528.

TMP (1820b). Mr Accum at home (Original poetry, six verses), *TMP*, Saturday, December 23, Issue 15529.

TT (1818). Court of King's Bench, Tuesday, *TT,* Wednesday, Dec 09, p. 3; Issue 10538.

TT (1819). Advertisement for Accum's chymical amusement, *TT*, Wednesday, Feb 10, p. 4, Issue 10593.

TT(1820). Advertisement for *A treatise on the art of brewing*, *TT,* Tuesday, Nov 28, p. 4; Issue 11103.

TT (1821a). Westminster Sessions, January 4, *TT,* Saturday, Jan 06, p. 3; Issue 11137.

TT (1821b). Mr Frederick Accum, *TT,* Wednesday, Jan 10, p. 3, Issue 11140.
TT (1821c). Westminster Sessions. Mr. Accum's Case, *TT*, Friday, Apr 06, p. 3; Issue 11213.
TT (1821d). Sales by auction, *TT,* Monday, Apr 16, p. 4; Issue 11221.

ABBREVIATIONS FOR NEWSPAPERS

BNP The Bury and Norwich Post: Or Suffolk, Essex, Cambridge, Ely, and Norfolk Telegraph. (Norwich)
CM *Caledonian Mercury* (Edinburgh)
GH *Glasgow Herald* (Glasgow)
RCG *Royal Cornwall Gazette, Falmouth Packet & Plymouth Journal,* (Truro)
TLM *The Leeds Mercury,* (Leeds)
TMC The Morning Chronicle (London)
TMP *The Morning Post* (London)
TT *The Times* (London)

In: New Developments in Science Education Research
Editor: Nathan L. Yates
ISBN: 978-1-63463-914-9

Chapter 7

THE EXHIBITIONS OF THE NATURAL HISTORY PALEONTOLOGY AREA AT THE BRITISH NATURAL HISTORY MUSEUM BASED ON ARGUMENTATION

Jun-Young Oh*[1] and Dong Joo Han[2]
[1]Center for Integrated General Education, Hanyang University, Seoul, Republic of Korea
[2]Department of Astronomy & Space Science, College of Natural Science, Chungbuk National University, Cheonju, Republic of Korea

ABSTRACT

The aims of this research are to examine an argumentation approach to the layouts of exhibitions and the structure of explanations used in one of the British Natural History Museum's Informal educational institutions. For this purpose, we proceed in accordance with the research questions: first, we have established Checklists for evaluating the criteria of macro and micro perspectives of the layouts of exhibitions and the structure of explanations based on argument. The arguments applied to this research, which focuses on the Natural History Museum's Dinosaur Hall, are as follows: (a) Booth et al.'s (2008) argument for problem-solving and Problem-proposing, as well as (b) Oh's (2012, 2014) scientific reasoning argument, for the scientific methodology, were well applied to Natural History Museum Dinosaur Hall.

Keywords: Argument, natural history museum, exhibitions, scientific reasoning argument, interactive exhibit

* Jyoh3324@hanyang.ac.kr.

Introduction

According to Conn (2011), these are the expectations that visitors bring with them when they enter a science museum:

> Science museums would present the world understood, organized, and managed, and so doing reinforce the very idea of the power of science. Further, those expectations were, I believe, because museums conveyed these messages by means of objects. Documents, and the people who wrote them, after all, might lie, but museum specimens did not. These beliefs, modified across the twentieth century to be sure, most likely still shape the expectations of many who visit science museums. They still expect the museum to present exhibits that demonstrate science's conclusiveness, rather than its doubt, and that portray science as a central part of our progressive world. That science museums could display controversial ideas is not merely offensive to some, it fundamentally contradicts the very understanding many have about what "science" is in the first place (Conn, 2011, p. 496).

Conn (2011) states that traditionally, science education in a Science Museum is taught primarily through convergent activities focusing on a correct answer, rather than the process of continuous exploration, which is said to be the true form of science. However, students are typically unable to perform scientific activities when the correct answer is unknown, unavailable or can be explained through various methods or when they can express their opinions. Because students do not understand why these methods are necessary, they ultimately do not associate hands-on activities with thinking activities with regard to scientific concepts (Hodson, 1996; White, 1996; Walton, 2000).

Consistent with the notion that studying science means studying the culture of scientists, 'authentic science' can be discovered by the students who participate in scientific activities with interest and enjoyment.

To solve this problem, a variety of learning resources and opportunities should be created using diverse learning environments, including a shift in educational focus from classrooms to the outside world, so that the students can recognize issues related to natural phenomena and make real discoveries in their real-world environments.

As the importance of the educational function of science museums as informal educational institutions grows, the scientific basis for the assessment and advancement of exhibit layouts and descriptive texts presented within science museums should be examined so that the exhibits' induction of creative thinking can be evaluated. The efficacy of storytelling methods and compositions, which are significant to scientific education, should be investigated through an evaluation of learning styles with regard to the story composition of an exhibit and the content development method. One important function of an argument is that the speaker convinces his readers or listeners of the value of his claims using elements of his argument (Oh, & Kim, 2009). Therefore, it is necessary to investigate whether exhibits in science museums, as informal resources rather than formal materials used in classroom education, provide stories with meaningful content that is compelling to visitors.

Hein (1998) argues that discovery learning in science museums requires active learning situations to provide learners with opportunities to operate experience and test ideas and made the following statement regarding active learning and the arrangement of exhibitions:

> Active learning is often translated into physical activity associated with learning; thus the common reference to "hands-on" learning. ... But ... Monotonous repetitive physical activity, or "mindless" actions are not particularly conducive to mental changes associated with form of learning, and recent literature has stressed the need for "minds-on" as well as "hands-on" engagement by learners.... Such discovery rooms would fall under the heading of constructivist education facilities. (Hein, 1998, pp. 30-31).
>
> Exhibits in a gallery designed on discovery lines may or may not be arranged linearly, with a beginning and end. If the entire exhibition is intended to lead the visitors to appropriate conclusions, the placement of exhibit components may provide such guidance. (Hein, 1998, p.33).

According to Hein, the requirements for the use of discovery learning in the science museum are as follows: First, for active learning, he has stressed the need for "minds-on" as well as "hands-on" learner engagement. Second, if the curatorial staff's intention is for the arrangement of exhibits to convey the meaning of scientists, it is good for the beginning and the end to follow such a guide.

Therefore, as a specific example, the Dinosaur Exhibition at the Natural History Museum in London, which covers a wealth of past geological events and theory development, will be explored because Science museum exhibits display questions that have been answered and debates based on this discovery constructivist museum as two criteria.

The paleontology exhibit at the Natural History Museum encompasses extensive evidence and materials related to how past historical scientific evidence was handled and how theories developed. The primary purpose of this study is to find applications for classroom education by examining the scenario compositions and description structures of museum exhibits from both macroscopic and microscopic perspectives. Exhibits tend to use films and plays to develop a scenario, which often provides the opportunity to recognize and utilize the value of the material composition of the paleontology exhibit from a scientific and educational perspective. For this purpose, we will establish the following specific research problems to define our preliminary arguments and checklists for evaluative factors for the arguments that other research (Lee, et al., 2005) has not examined.

First, we developed these checklists using a general structural analysis method for exhibit scenarios according to elements of argument listed by Booth et al. (2008).

Second, in accordance with these checklists, in which argument structures, from macroscopic perspectives, apply to the paleontology exhibit theme and scenario composition at the British Natural History Museum?

Third, accordance with these checklists, in which argument structures, from the Intermediate theme and microscopic perspectives, apply to the paleontology exhibit theme and scenario composition at the British Natural History Museum? How can Oh's (2012; 2014) interactive-type scientific reasoning methods be adopted?

THEORETICAL BACKGROUNDS

For exhibition scenarios at natural history museums, which are considered informal scientific teaching and learning resources, Lee Sun-Kyung et al. (2005) developed a major theme from a macroscopic perspective with an introduction, development, and conclusion that corresponded to a dramatic scenario. In addition, the relationship between the intermediate theme and minor theme was analyzed from a microscopic perspective using Toulmin's argument structure, Magnani's (2001, 2009) S-T argumentation activity, and Oh's (2012) Natural Enquiry Model, with the British Natural History Museum being used as an example.

Oh (2012, p.231) claimed that unlike the hypothetico-deductive method, which has traditionally been recognized as a scientific method that establishes and verifies a hypothesis, the hypothetico-deductive method involving abduction establishes several alternative hypotheses and uses the most reasonable one to explain a phenomenon. The use of the hypothetico-deductive method involving abduction plays an important role in the development of scientific hypotheses and theories. That is, alternative hypotheses are considered important, and it is evident that the abduction method, which is a description method that uses information obtained from familiar situations to describe a new situation, has been applied.

In this study, we introduce arguments of Booth et al. (2008). In addition, Oh's (2013) scientific reasoning method, which Peirce's abduction recently added, will be explored.

1. Booth et al.'s (2008) Argument As a Conversation with Readers

Similar to Craft's Revision of Toulmin, Williams et al. (2003) remove the arrows because most arguments begin not with a statement of grounds but with a problem. This approach lost backing because Toulmin required backing to explain how arguments differ among different fields. Backing refers to the support of a warrant, which is viewed as a claim in its own argument, and an eliminated qualifier is considered a distinct element because qualifiers are not a singular element of an argument like a claim or a reason; qualifiers color every element-claim, reason, piece of evidence, warrant, and rebuttal.

Similar to Craft's Revision of Toulmin, Williams et al. (2003) remove the arrows because most arguments begin not with a statement of grounds but with a problem lost support, because Toulmin required backing to explain how arguments differ among different fields, referring to the support a warrant, viewed as a claim in its own argument, and eliminated qualifier as a distinct element because qualifiers are not a singular element of an argument like a claim or a reason; qualifiers color every element-claim, reason, piece of evidence, warrant, and rebuttal.

Additionally, they divided the single element "grounds" into two parts, reason and evidence, because careful readers accept a claim regarding a contested issue only when there are two distinct types of support: reason and the evidence on which this reason rests. However, reasons provide only the logical structure of that support; evidence is the basis on which that structure of reasons rest and is brought in from outside the argument. They replaced rebuttal with acknowledgment and response because Toulmin defines rebuttals as limits on the scope of a claim, but in ordinary language, what we call a rebuttal responds to objections of any type—not only to the scope of a claim but also to the source or sufficiency of its support, to the soundness of logic, to the definition of a problem, and to alternative solutions. Thus, they expand Toulmin's rebuttal to refer to responses to any anticipated alternative, objection, or criticism. However, they believe that the term rebuttal can encourage responses that are too aggressive, so they provide substitutes that are more amiable and accurate: acknowledgment and response.

Supporting My Claim

At the core of every research report is the answer to the research question, the solution to the problem--- the primary *Claim*. One must support this claim with two types of support: reasons and evidence.

The first type of support, a reason, is a statement that provides my readers with cause to accept my claim. We often join a reason to a claim with *because.* I usually require more than one reason to support a contestable claim and, in a detailed argument, each reason will usually constitute a separate sentence.

The second type of support is the evidence on which I base my reasons.

With reasons and evidence, we have the core of a research argument:

CLAIM because of REASONS based on EVIDENCE

However, to offer a complete argument, I must add at least one additional element and, often, a second: I must acknowledge other points of view and offer what we call warrants, which explain how a reason is relevant to a claim.

Careful readers will question every part of my argument, so I must anticipate as many of their questions as I can and then acknowledge and respond to the most important ones.

Even when my readers agree that a reason is correct, they may still object that it is not relevant to my claim. We can add warrants to our diagram (Figure 1) to demonstrate that they connect a claim and its supporting reason:

I make a claim, support it with reasons, support them with evidence, acknowledge and respond to other views and, sometimes, explain my principles of reasoning.

These five elements constitute the core of a 'basic' argument. However, arguments in research reports are more complex (Booth et al., 2008, pp. 116-117; Williams and Colomb, 2003, p.118).

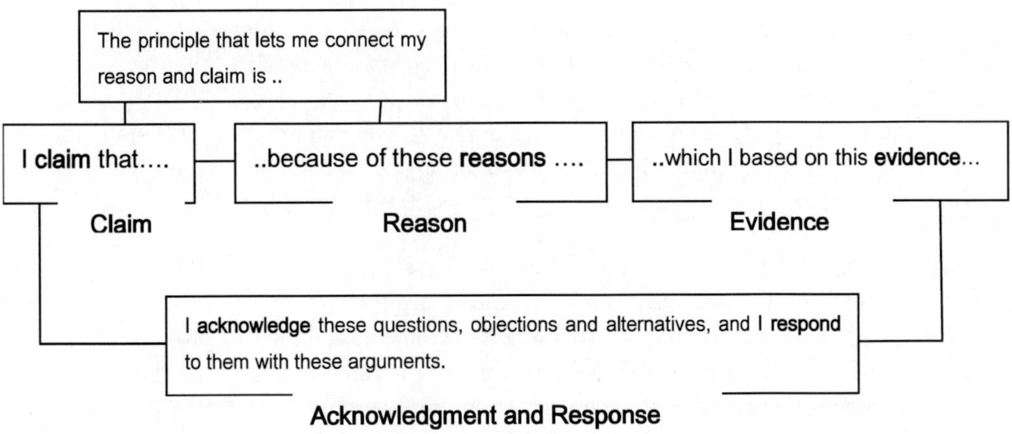

Figure 1. The elements of Booth's et al.'s (Booth, et al., 2008, p.35) argument.

First, we almost always support a claim with one or more reasons, each of which must be supported by its own evidence and perhaps justified by its own warrant, because readers may be reluctant to accept only one reason as sufficient support for a significant claim.

Reasons in Parallel: I can offer parallel reasons, each one directly supporting a claim.

Reasons in Sequence: I can offer stacked reasons, each one resting on the one before, the first one directly supporting my claim, and the last one resting on evidence (or a report of it).

The Deep Complexity of Serious Arguments: when we address an issue that requires more than two or three pages of argument, we build our case using multiple reasons in parallel, each of which rests on multiple reasons in sequence, all of them based on evidence.

Second, because readers think of many alternatives and objections to any complex argument, careful researchers typically must respond to more than one or two of them.

Third, moreover, each element of an argument may itself have to be treated as a subclaim, supported by its own argument:

Each response to an objection may require reasons and evidence to support it.

If my readers doubt the truth of a warrant, I may have to treat it as a subclaim and support it with its own argument, including reasons, evidence and, perhaps, even its own warrant with its own acknowledgments and responses.

2. Basic Inferences of Scientific Reasoning in terms of Abduction, Involving the Deduction-Induction Method using Argument (Magnani, 2001, p. 74; Magnani, 2009, p.18; Oh, 2012, p.231; Oh, 2014)

Hein stated that an idealist curator might more likely consider displaying multiple perspectives (Hein, 1998):

> The important point for this argument is that in a realist epistemology, the focus of the museum content is guided by material being displayed, by *the nature of the subject*. It is not organized with a primary concern for the viewer's interest, or the meaning the viewer might make from material. In contrast, the idealist curator believes that meaning of an object (or an entire exhibition) derives not from some external reality, but arises from the interpretation it is given, either by the curator or by the viewer. Thus, a curator who ascribes to an epistemology towards the right of the continuum (idealist) might more likely consider showing multiple perspectives, or arranging an exhibition so that it allows visitors to draw various conclusions from their interactions with it (Hein, 1998, p. 21).

However, convincing others of the validity of causal claims is certainly an important component of science. The present view of scientific argumentation differs somewhat from that proposed by Toulmin in that it emphasizes argumentation in the process of testing alternative hypotheses. In other words, the initial goal of argumentation in the testing process is to discover which of two or more proposed alternative explanations (claims) regarding a puzzling observation is correct and which of the alternatives are incorrect. In this sense, a claim can be thought of as a tentative explanation that may be correct, thus requiring testing through the generation of specific predictions and the gathering of evidence. Of course, once Scientists or students reach a sound conclusion (i.e. have 'discovered' the 'correct' explanation), they must once engage in argumentation to convince others of the veracity of their conclusion. In theory, this process should amount to constructing arguments for one of the alternatives and against the others (Lawson, 2003). Nevertheless, in both Booth et al.'s view (2008) and the present view, scientific arguments contain claims and reasons and may also contain evidence and warrants.

Using this argument as an example, figure 3 depicts the major elements of Deduction-Induction argumentation and the key words that link the elements. The perceptible elements appear as 'solid boxes' in the environment region, whereas the imagined elements appear as 'clouds' in the mind region.

Abduction-Retroduction Phase

As shown, the process begins with a puzzling observation that provokes a causal question and then the generation of one or more tentative explanations (alternative hypotheses-alternatives that often parallel explanations generated by previous generations of scientists; cf. Wandersee et al. 1994). As will be discussed, hypothesis generation is a creative, open-ended activity. Thus, the list of generated hypothesis may in fact not contain a hypothesis that may eventually be found 'correct'.

We understand a new exploration that leads to a puzzling observation.

Then, because of our prior store of declarative knowledge, we use abduction to generate possible hypotheses (tentative explanations).

Next, we use retroduction to subconsciously test our hypotheses, which lead to some doubt and then to rejection, or which, when presumably tested using retroduction, are supported; thus, we use retroduction to select and rank hypotheses.

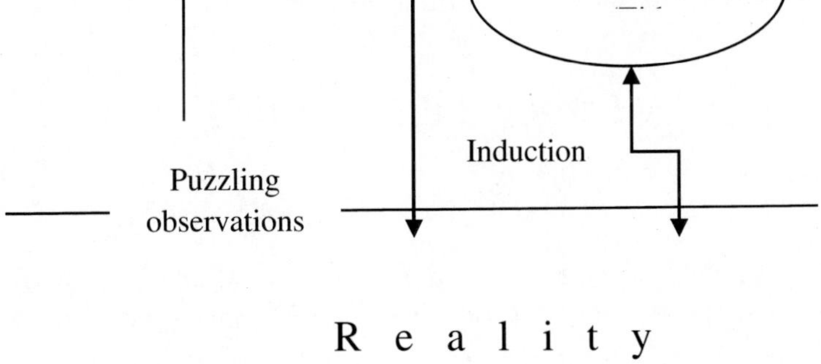

Figure 2. The four stages of Scientific Inquiry involving Retroduction [reformed version of Flash and Kakas, (2000, p.7)].

Once hypotheses have been selected, they must be ranked to plan the evaluation phase by first using a certain preferred hypothesis (Magnani, 2001, p.73). Arguments for the best explanation cite a range of explained facts. However, there is also a dynamic notion of consilience that must be contemplated when considering the acceptability of explanatory hypotheses. Explanatory criteria are needed because the rejection of a hypothesis requires demonstration that a competing hypothesis provides a better explanation. Clearly, in some cases – for instance, when choosing scientific hypotheses or theories, in which the role of "explanation" is dominant – conclusions are reached in accordance with rational criteria such as consilience or simplicity (Magnani 2001, p.27). The "Selection of hypothesis" covers all of the phenomena that are present; next, the first preliminary test stage occurs, in which tests are planned to select or eliminate hypotheses using retroductive strategies, which is a weak test of a hypothesis in the sense that the least that a hypothesis should do is explain the puzzling observation that led to its generation in the first place using what we already know (Lawson 2010).

Deduction-Induction Phase

Regardless of this limitation, once generated, the hypotheses must be tested. To test a hypothesis, one begins by assuming that the hypothesis is correct. Next, one must imagine a test that, with the hypothesis, should produce one or more specific observable results (the prediction). The words If/and/then link the hypothesis and the imagined/planned test to prediction.

Once a test is planned, complete with one or more predictions, the test is conducted. The test then yields observed results. The observed results constitute evidence. Finally, the evidence is compared with the prediction. The match or mismatch between evidence and prediction is then used to draw a conclusion regarding the degree of support or non-support for the hypothesis.

1. We then used deduction to generate future predictions, which also require connections to declarative knowledge.
2. Subsequently, we made the necessary observations, which matched our predictions.
3. Finally, on the basis of this match, we used induction to draw the conclusion that is supported.

METHOD AND RESULTS

1. Subjects

The subjects of this study were the structure and contents of an exhibition scenario for a dinosaur exhibition hall at the Natural History Museum in London. One of the world's most famous museums, the Natural History Museum in London is interactive, with Exhibition, Education, and Research being the primary functions of the museum. Dinosaur Exhibition Hall is the main part of the Natural History Museum in London. Specifically, Dinosaur investigations require knowledge of natural science, as well as the use of a natural science enquiry procedure using problem-solving as a type of instruction-learning material in informal science learning.

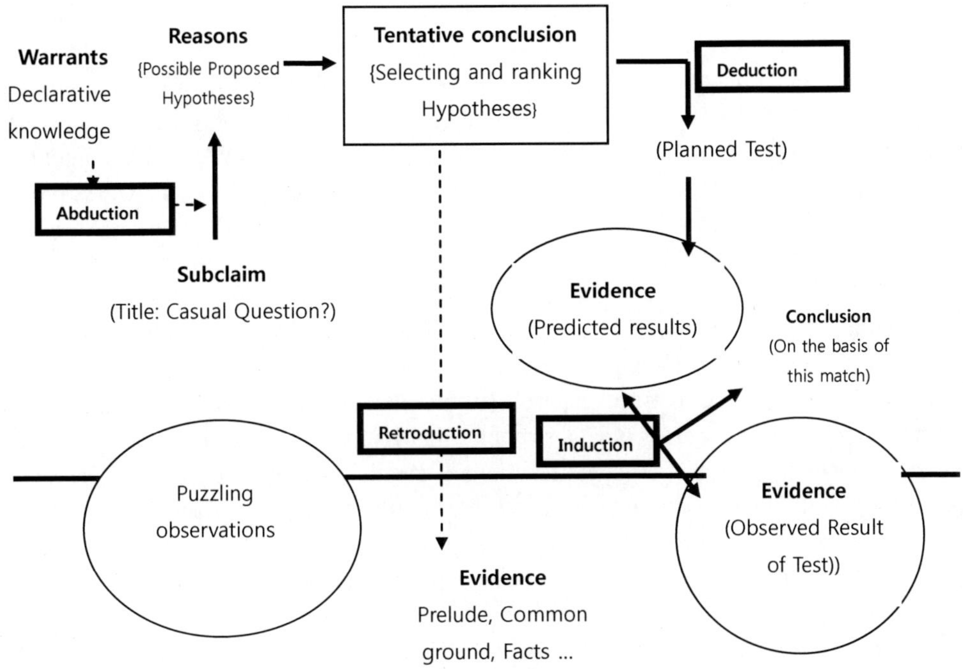

Figure 3. Deduction-Induction cycle involving Abduction-Retroduction Using argument.

2. Data Collection and Data Analysis

Two researchers participated in this study, photographing the Dinosaurs Exhibition in its entirety, and collecting all materials for explanatory purposes and web-materials through investigation of the web-site for the Dinosaurs Exhibition; we visited the museum from 15 March to 16 March, 2011.

The analysis of data for the museum exhibition materials and Explanatory statements included three steps.

First step: we developed a Typical Analysis Method for Exhibition Scenario based on Booth et al. (2008).

Second step: we analyzed the macroscopic perspective (Introduction- Development-Conclusion) of an exhibit scenario for an argument structure at the British Natural History Museum based on Booth's argumentation.

Third step: we selected the exemplary for an Intermediate theme and Elementary theme for a Display Model of Exhibition based on Booth's Argumentation and for an Interactive model of exhibition based on Magnani's (2001), and Oh's (2012; 2014) Hypothetico-Induction involving abduction.

As the role of science museums develops with regard to social education, in addition to being places where the general public can view exhibits with interest, science museums have extended their role, transforming into spaces where visitors can acquire knowledge of natural sciences. In recent years, nature has been viewed as both a way of living for humans and a survival environment. Both modern science and historical science have explored the coexistence of human beings and nature through the proper recognition of nature. Thus, both taxonomic exhibitions and various natural phenomena have been established as themes in science museums to evoke visitors' attention. Expectations for science museums to serve as institutions for social educational knowledge began to grow as well. To solve such problems, scenarios from other fields should be compared, and their positive outcomes should be adopted. For example, dramatic scenarios from theatrical films or plays provide contexts that are similar to exhibition scenarios at science museums in that the production process typically includes creative works through the use of various media and technology. In addition, large amounts of capital and the ability to direct and edit are considered important in both scenarios. A primary difference between scenarios is that a play scenario has a focus on expression through dialogue and a film scenario has a focus on expression using images, whereas exhibition scenarios tend not to focus on these factors.

Despite these differences, film scenarios are similar to exhibit scenarios with regard to structural aspects. Film scenarios have a structure that includes an introduction, development, turn, and conclusion. Using a specific theme, events progress along this descriptive framework through the use of various expressions. Finally, the event delivers a message to the audience. In this sense, film scenarios can be used to structure exhibit scenarios. The application of the structure of dramatic scenarios provides a story-like quality in an exhibit space in that the possibility of evoking anticipation regarding another space becomes inherent when moving from space to space. To maximize the purpose and effectiveness of an exhibit centered on a specific theme, the entire exhibit space should be organized internally into an introduction, body, and conclusion from a macroscopic perspective. Additionally, an argument structure should be consistently provided between each individual display and description with regard to context from microscopic and macroscopic perspectives. Therefore, the macroscopic structure of an exhibit scenario and microscopic inspection standards for an argument structure were investigated at the British Natural History Museum.

First Step

Typical Analysis Method for Exhibition Scenario Based on Booth et al. (2008)

Macroscopic Perspective

Although the order of elements in a written argument can vary, some standard order can be relied upon: The default order for the three core elements is Claim + Reason + Evidence. Because a serious claim can rarely be supported with only one reason, core elements are arranged by simply adding more Reason + Evidence pairs. These elements are sandwiched between an introduction that states the problem and solution (the primary claim) and a conclusion that restates that claim (Williams & Colomb, 2007, p.47).

First, A prelude is best used when it introduces key themes that an argument will pursue, particularly when it represents the problem with a vivid and concrete example. The structure of the fullest introduction will now be described (Williams, et al., 2003,p. 74).

When we reason regarding a problem, we do not begin with grounds and then think our way to a claim (its solution). We begin with the problem that motivates us to search for a solution in the first place and find a tentative hypothesis based on the facts that are then available to us. We then use the hypothesis (C. S. Peirce calls it a "hypothesis on probation") to find more data that we hope will confirm or disconfirm it (Wiliiams and Colomb 2007, pxxv).

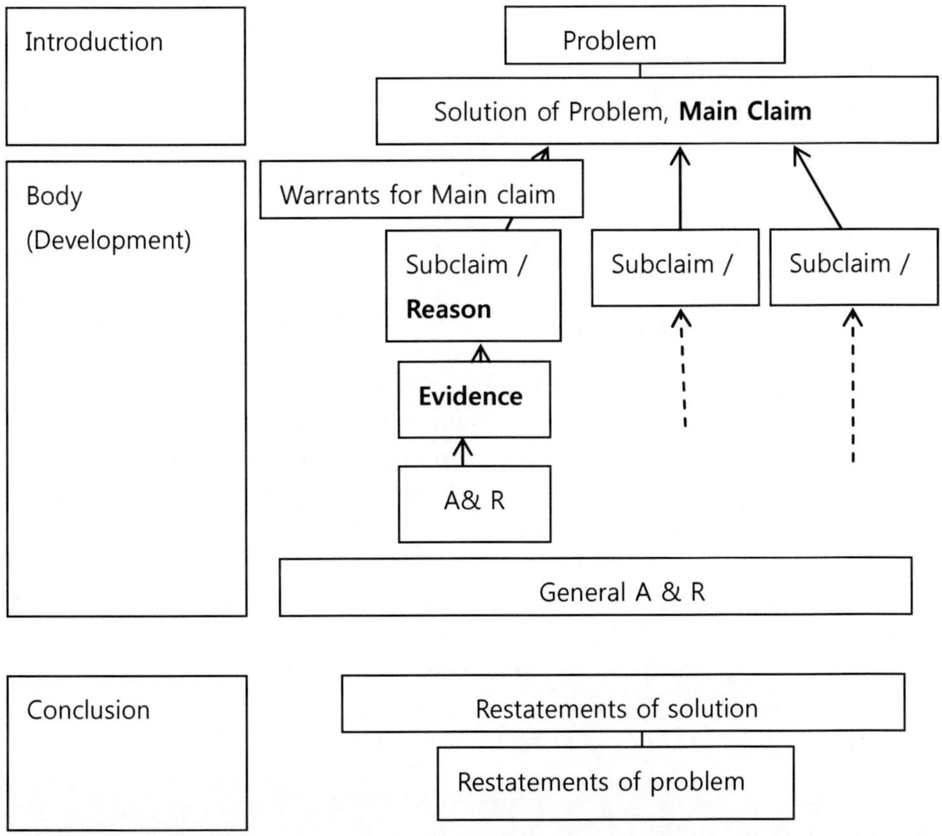

Figure 4. Default Plan for the Core of a Modified Complete Argument (Williams and Colomb, 2007, p.48).

The introduction begins with a motivating issue so that visitors can understand the basic framework of the entire exhibit. The best way to easily understand this basic framework is to call attention to the inherent familiarity that exists between exhibit contents and visitors. Such familiarity can easily be found through the use of everyday terms for explanations, as well as by making a connection between exhibit contents and daily life. The introduction first introduces the issue being investigated and then addresses future directions for development.

Second, Because any serious claim will rarely be supported with only one reason, core elements are arranged by simply adding more Reason + Evidence pairs. These elements (the body) are sandwiched between an introduction that states the problem and solution (the primary claim) and a conclusion that restates that claim (Wiliiams and Colomb 2007, p. 47).

These elements are almost always omitted when the readers share your value, assumptions, definitions, and so on. If they do not, you may have to state them as warrants, typically before the reasons to which they apply and, perhaps, even support them with their own reasons and evidence (Williams and Colomb, 2003, p.43).

Third, their parts can be inversely mapped onto the parts of the introduction: first, begin the conclusion by restating the gist of the primary claim; second, explain its significance by answering *So What?*; third, suggest an additional question or problem to be resolved, that is, something that is still not known, and end with an anecdote, question, or fact that echoes your prelude (Williams, and Colomb., 2003, p.76).

The conclusion should summarize the exhibit contents, should be provided in abstract terms with regard to the specific exhibit contents and should encourage a continued desire for exploration.

Microscopic Perspective

Exhibition scenarios are explored using the argument structure of Booth et al. (2008) and Lawson within the context of macroscopic and microscopic perspectives.

As illustrated in Figure 5, any reasoning inferred from the macroscopic perspective in Figure 5 – that is, reasoning regarding claim, reason or evidence – progresses and is developed as another sub-claim (Booth et al., 2008, p.116). However, any element of argument can be used as a sub-claim to develop the argument. Thus, microscopic structure can be considered a sub-element of macroscopic structure. Therefore, the viewpoint involving the justification of claims and reasons from a microscopic perspective follows the inspection criterion of development, which perceives introduction from a macroscopic perspective and develops the argument. The manners in which arguments are developed from macroscopic and microscopic perspectives are similar in that the basis of scientific argument is the justification of one's claim. Although scientific principles, as general concepts, can be presented initially, these principles can also diminish interest. Therefore, scientific principles should be identified following provision of the reason for the exhibit or at the conclusion of the exhibit or even presented as the solution to the problem.

Justification of Claims and Reasons: Display Model of Exhibition

Checklists for Display Model of Exhibition

Check of argument structure in a macroscopic context. Traditional museum exhibits were organized in an elegant and quiet atmosphere. Exhibits were viewed through glass cases, and the museums utilized moving models. These methods are known as Hands-off Exhibitions. Currently, these hands-off methods cannot compete with flashy and over-stimulating pop culture. Therefore, it is necessary to use an argument approach to actively encourage audiences to visit exhibits.

To justify the claims of the exhibit, the solution to the problem should be restated in the conclusion within the general form of the exhibit. In addition, the warrant, which is a scientific principle that connects claims and reasons, could be provided in the beginning as a general concept or could be presented as a premise for the conclusion. Argument strategies for claim justifications can largely be categorized as problem-solving arguments and problem-posing arguments (Williams and Colomb, 2003, pp.75-76). If a claim justification is represented as a problem-solving argument, this strategy makes the solution to the problem evident by lining up the reasons in a serial arrangement as much as possible.

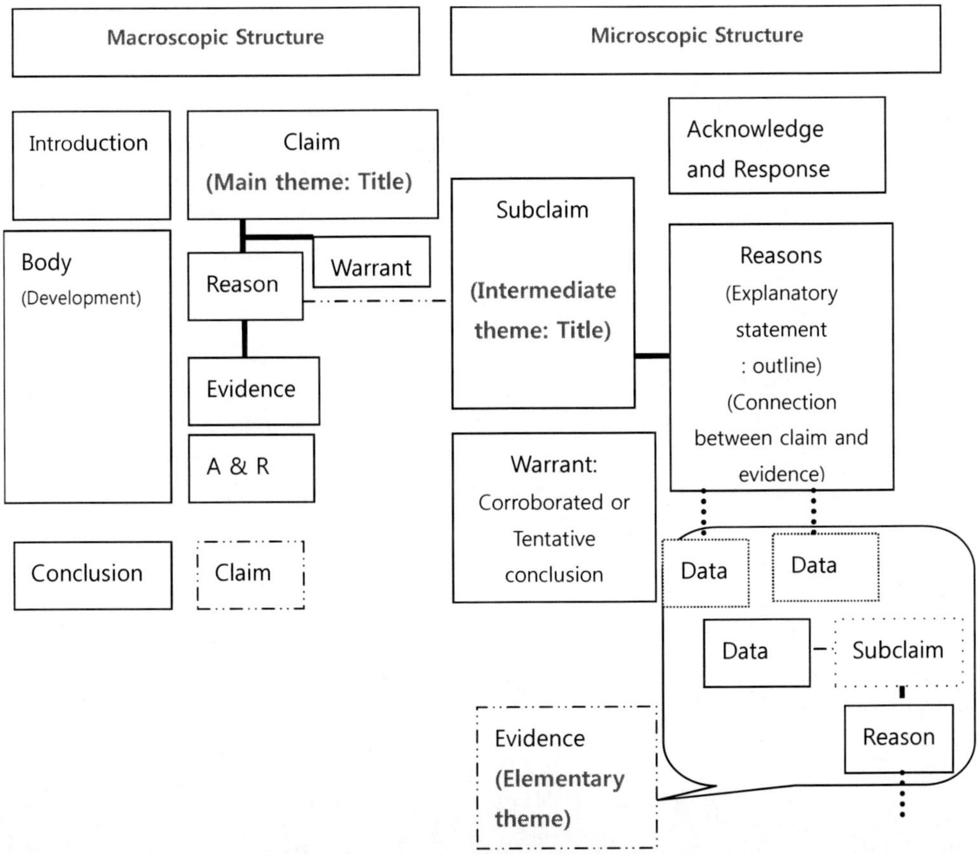

Figure 5. The argument elements of claim justification modified using Booth et al.'s (2008) argument.

Interactive Exhibits (Hands on Behavior) in Science Museums: Oh (2012; 2014)'s Argumentation for a Practical Enquiry Behavior

Interactivity is viewed as a vital characteristic of science museums (Allen, 2004). Visitors spend more time at interactive exhibits than at non-interactive exhibits, and interactivity is associated with superior learning and recall of information (Allen, 2004; Sandifer, 2003). The effects of different aspects of interactivity on visitor behavior and learning have been studied (e.g., Afonso & Gilbert, 2007; Sandifer, 2003). For example, Sandifer found that open-endedness and technological novelty influence the amount of time that visitors spend at exhibits. Interactivity is an important ingredient in adult visitors' science museum experience, but it is an indispensable part of young children's visits. Hands-on experiences are considered crucial for science learning in this group (French, 2004; Gelman & Brenneman, 2004). The emphasis on interactivity with respect to young children calls for a strong need to investigate this age group's on-exhibit behavior (Van Schijndel et al., 2010).

Interactive exhibits, which are also known as Hands-On Exhibits, prioritize an individual's hands-on experience in the exhibit. Unlike traditional science museums with exhibits displayed in glass showcases, interactive exhibits can be touched, and audiences can learn by being more proactive. One of the benefits of interactive exhibits is that they are designed to provide opportunities for audiences to discover something on their own. The key point of such exhibits is to induce the audience to enhance their own cognitive abilities. Voluntary, not forced, natural learning through enjoyable plays instead of implanted knowledge leaves a deep, lasting impression on audiences. Generally, when hands-on exhibits are examined, they are classified as direct or indirect experiences. Direct experience refers to participation using parts of the body, whereas indirect experience refers to the method of assisting understanding through the use of presentation methods, such as display media.

The following argument approach is necessary for hands-on exhibits: the warrant, which consists primarily of scientific principles, should be provided at the beginning of the exhibit as a general concept. However, once a test is planned, more than one prediction should be made. The test will be performed and will yield a result, which constitutes evidence. Ultimately, this evidence will be compared with the prediction. Hypothesis-predictive arguments can also largely be divided into problem-solving arguments and problem-posing arguments. One hands-on argument, shown in Figure 6, is a type of problem-solving argument for a case that presents the solution clearly in the conclusion (e.g., corroborated solution). In contrast, a conclusion in a problem-posing argument can produce several solutions given that the crucial experiment was not performed. Therefore, these conclusions can be expressed as provisional conclusions in that they suggest a solution (Williams and Colomb, 2003, p.75). Based on the views of Lawson (2003, 2010) and Oh (2012; 2014) regarding hypothetico-deductive argumentation, the following checklists for a hands-on exhibit were established.

Second Step

Interpretation of Exhibit Scenarios at the Paleontology Exhibition in the British Natural History Museum

When the paleontology exhibition scenario was contrasted with the composition of a scenario for a play, a structure including an exposition, development and a conclusion was evident with regard to the development of the exhibition contents. Here, we will use both the macroscopic and the microscopic perspectives to determine what type of argument structure the scenario of the Paleontology Exhibition at the British Natural History Museum used by reorganizing according to argument structure.

Development of the Exhibit Scenario from a Macroscopic Perspective

<Introduction of the Exhibit Scenario >

The intermediate theme, which was 'exposition' for the paleontology exhibition, was composed of several things. One of the intermediate themes was the following: 'What kind of creatures were dinosaurs?' The survival periods and physical features of dinosaurs were available for view via panels and touch screens.

■ ***Checklists for Introduction: Understanding the basic framework of the entire exhibition***

1. Would you supplement your main claim/problem with a prelude, quotation, interesting fact, or brief anecdote that encapsulates the problem?

The explanations included the following information: dinosaurs were reptiles; they lived 65 to 230 million years ago, they were land animals that could not fly, and they walked with straight legs. Based on this information, visitors can answer quiz questions on dinosaurs using a touch screen. This information also provides hints for the visitors with regard to the direction in which the exhibition is headed and allows their anticipation to build regarding the minor theme.

2. If your problem is conceptual, have you stated clearly what is not known or not sufficiently understood? Have you stated the consequences in a way that shows that they are more significant than the destabilizing condition?

An intermediate theme, which is the introduction, is available in the form of a question that encourages visitor interest in the exhibit's future contents. A quiz provides visitors with both background knowledge regarding the exhibit's future and clues regarding the importance of the subject.

3. Where have you placed your main claim/solution?

The introduction has an argument structure that presents the main problem in question form and later reveals the main claim.

<Body or Development>

Responses to the following questions posed in the 'Introduction,' are developed and answered in various ways: 'what are dinosaurs?', 'how did dinosaurs adapt to the environment at the time?', 'what biological characteristics of dinosaurs are similar to those of extinct animals, and what are different?', and 'why were dinosaurs unable to adapt to their changing environment and therefore become extinct?'

■ ***Checklists for Body: Development of claims and argument activities for the question suggested***

1. You primarily omit them when your readers share your values, assumptions, definitions, principles and so on. If they are not shared, would you have to state them as warrants, typically *before stating the reasons* to which they apply?

The part considered 'Development' primarily addresses the biological characteristics of dinosaurs in various and extensive ways. It is developed through the use of information obtained from statistical data, verified statements obtained from another person, historical facts, examples, surveys, books or newspapers, and the Internet. Reasons and data available in the argument activity refer to all of the content that was offered to support one's claim. Therefore, development, which consists of these types of information, forms an argument structure that provides the answer to the claim in various ways.

2. Why have you ordered the reasons of your arguments as you have?

If you have divided the reasons for your argument into two or more parallel parts, can you explain their order?

If you have divided the reasons for the argument into two or more sequential parts, can you explain their order?

3. Locate acknowledgments and responses regarding which you suspect that visitors will think of objections and questions. Alternatively, list them all after providing the core of your argument (Williams and Colomb, 2007, p.48)

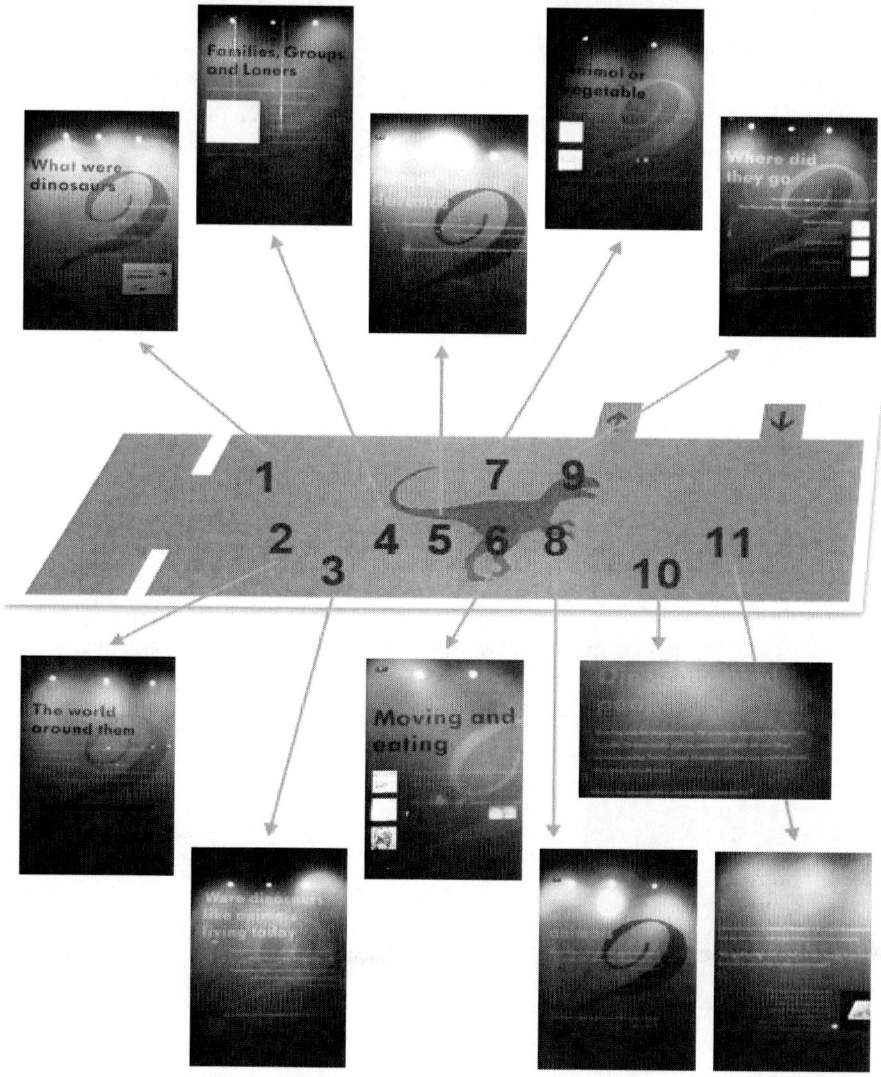

Figure 6. The exhibition space structure of Dinosaur Hall.

In the body, claims and argument activities are developed with regard to the proposed problem. Claims are completed by constructing and organizing appropriate arguments according to a given theme, which is followed by the qualified process.

<*Conclusion*>

In the conclusion, the question from the 'introduction' is concluded with the reasoning that it is because of human curiosity regarding disappeared animals and because of the scientists' efforts that the solution could be presented during the 'Development'.

■ ***Checklists for the Conclusion: Summary of the exhibition contents***

1. Have you restated the crux of your claim/solution at the beginning of your conclusions?

To develop a thorough understanding of the content, various human senses and those of dinosaurs are connected and organized sufficiently through the scientific process.

2. Have you suggested why the claim is significant?

Provide a full historical explanation of the efforts of the paleontologists who seek to study dinosaurs.

3. Have you suggested what is still unknown, uncertain, or unexplored?

Suggest that many species are becoming extinct even today and make visitors aware of the possibility that even humans may become extinct as dinosaurs did. In addition, do not forget the efforts of numerous scientists.

Reconstruction of the Macroscopic Exhibit Scenario by Theme

Figure 7 presents the authors' reconstruction of the overall data analysis as previously described according to the flow of the argument structure.

Third Step

Contents and Interpretation of the Explanatory Statements of the Exhibit from a Microscopic Perspective

From a macroscopic perspective, the argument structure has an introduction, development (Body) and conclusion. From a microscopic perspective, the intermediate and Elementary themes are classified within each Main theme. The exhibit scenario consists of one argument structure. Therefore, we will explore the argument structure of the exhibit scenario with regard to the intermediate and elementary themes of the Paleontology Exhibition at the British Natural History Museum.

Justification for Claims and Reasons: The Display Model of the Exhibition

Each middle theme, which corresponds to the 'development (body)', consisted of 3-6 minor themes. The display for each elementary theme explained the contents using commentaries and a variety of exhibition media. In most cases, individual elementary themes were gathered to form one consolidated intermediate theme. At this point, the exhibit explanatory statements consisted of the reasons for fossil evidence.

■ ***Questions regarding your Claims and Reasoning: an Inspection of the Display Model of the Exhibition***

1. Is your claim adequately significant to formulate an argument for it?

One of the intermediate themes was 'Were dinosaurs similar to extinct animals?' The use of the term 'living animal' and the claim that dinosaurs were living creatures similar to animals living today are important.

2. Have you maintained an open mind with regard to alternatives (i.e., explanatory statements)?

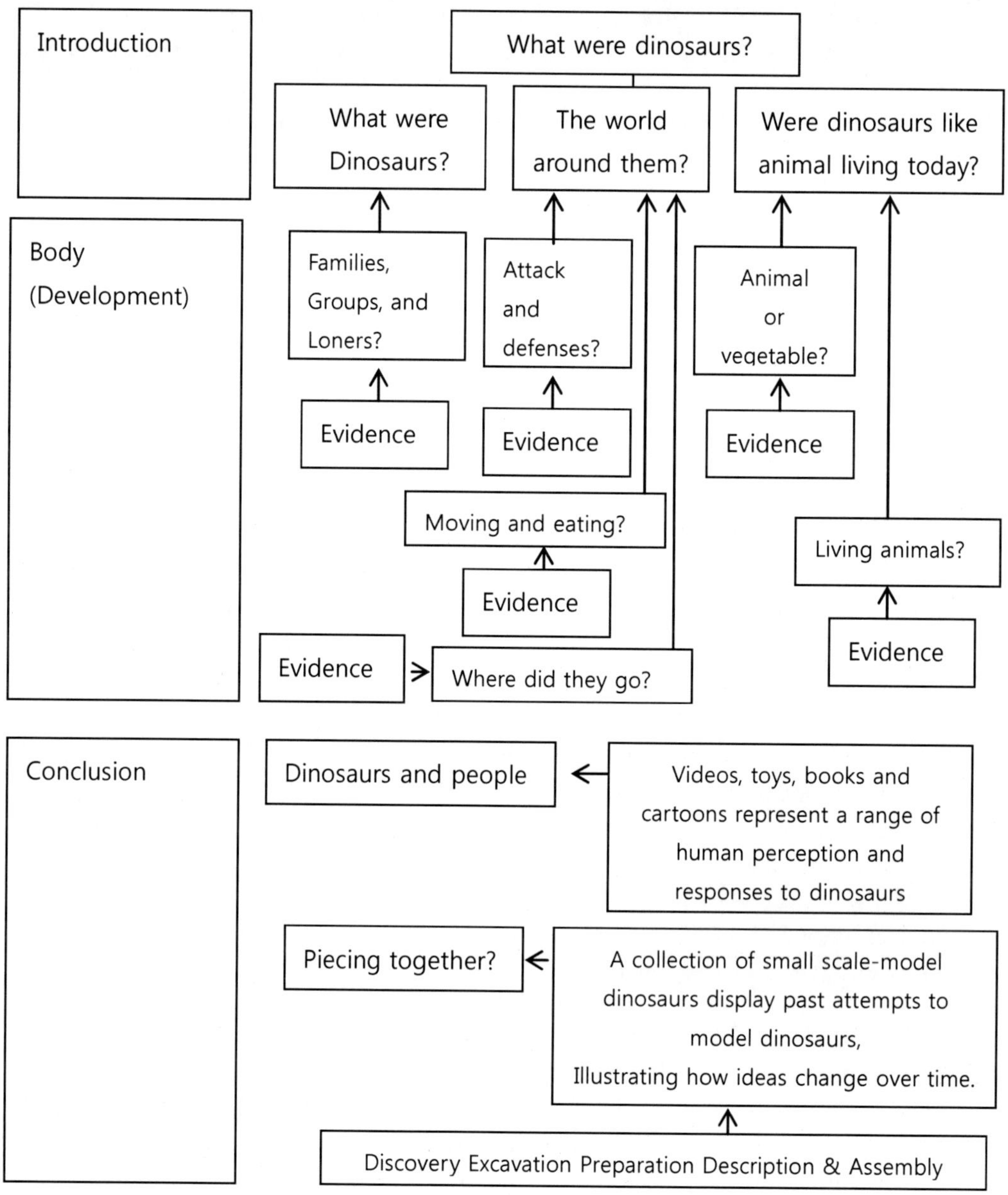

Figure 7. The reconstruction of the theme exhibition scenario.

3. Why have you ordered the reasons of your arguments as you have?

Explanatory statements, which correspond to the reasons for supporting the main claim for the intermediate theme, are expressed in the following words, which serve as the minimum grounds for the claim "on the basis of evidence presented, including the skeletons, legs, muscles and circulatory system of dinosaurs, the appearance and sizes of dinosaurs vary greatly, but they display a structure that is similar to that of birds, reptiles, and mammals." This structure is parallel, illustrating the appearance and size of and similarities between two things, but the basis of the main claim is their similarities.

■ ***Questions regarding your evidence (Elementary theme)***

1. Have you based your reasons on reliable evidence? Are your sources authoritative? Have you cited sources?

2. Do you possess sufficient evidence? Is the evidence representative?

The question, 'Given these similarities, can we consider dinosaurs to have the same characteristics as animals existing today?' (Can we use these similarities to understand dinosaurs as living animals?), suggests that there is a necessity for actual evidence supporting the explanatory statements, which corresponds to the reasons for supporting the claim for the intermediate theme.

Visitors are encouraged to continue questioning the claim. Evidence that provides a strong basis for the explanatory statements is provided by the role of elementary themes. The bones, skin, blood vessels, sensory organs, muscles, and lifespan that constitute the body structures of dinosaurs are contemplated in detail, as this evidence supports the 'reason' that dinosaurs had body structures similar to those of other animals.

■ ***Questions regarding your warrants***

Have you made explicit what your readers must believe before they will consider your particular claims and reasons (Williams, and Colomb, 2007, p.354)? The argument structure of the elementary themes consists of data and an overall theoretical basis that constitute a warrant. Concrete evidence and fossil samples are used as the representative evidence. As for a warrant, these common structures, including the spine, ribs, and straight-legs, are observed in the evolution of a wide variety of animal species.

■ ***Questions regarding acknowledgment and response***

Can you imagine a reader offering counter-reasons, counterevidence, or counter-analogies in response to your own? Can you imagine responding to them?

Exploration of Claims and Their Reasons: A Hands-On Model of an Exhibit

Hein (1998) claimed that a Museum must be organized around discovery learning. For example, a common idea invoked by advocates of active learning is that expressed in the Chinese proverb about teaching someone to fish:

Give me a fish and I eat for a day
Teach me to fish and I'll eat forever.

> The term "Discovery Learning" has been applied to all of the education ……, it has been used as a synonym for any form of education that attributes active participation to the learner. Many science and youth museums include a space called a discovery room, usually a resource room filled with materials of all sorts that visitors can explore and examine, often under the guidance of a staff member. Usually the emphasis is on exploration – asking questions and being able to have experiences with materials – rather than on arriving at conclusions or learning something in particular. In terms of our definitions, such discovery rooms would fall under the heading of constructivist education facilities (Hein, 1998, p. 31).

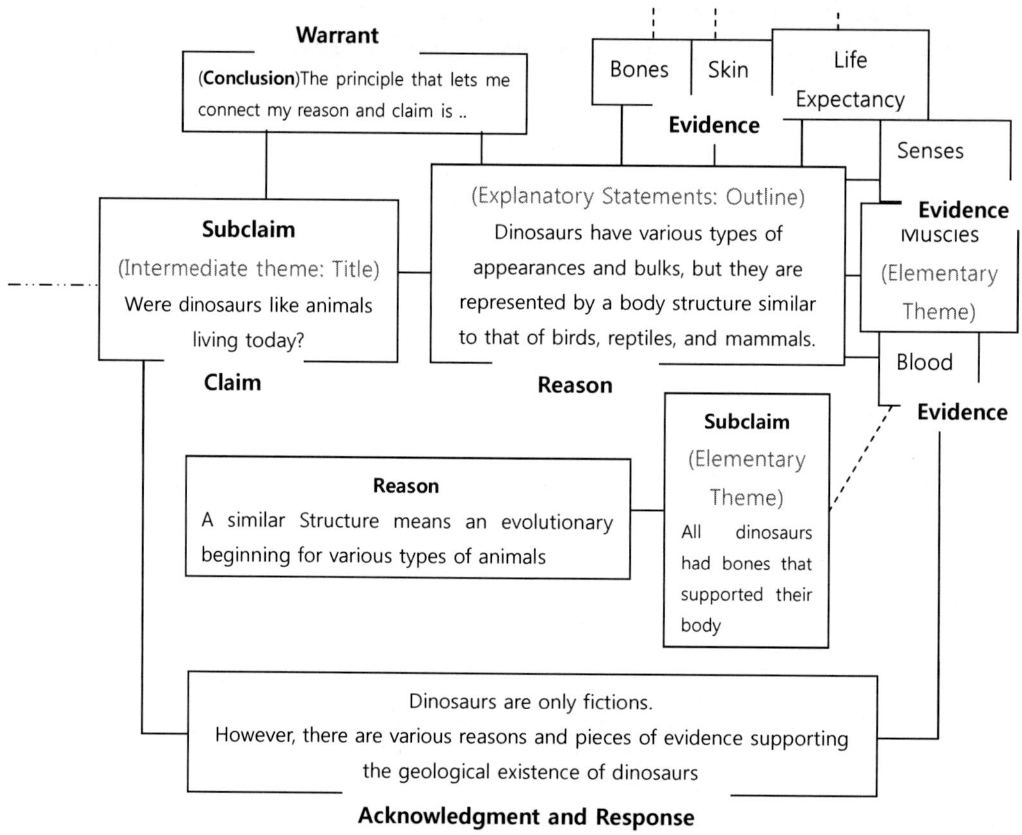

Figure 8. The justification argument for the claim and reasons that dinosaurs are similar to present animals.

Visitors may have several lingering questions regarding dinosaurs. For example, why did dinosaurs flourish during the period in which they did? Scientists provide reasons why dinosaurs were poikilothermal/homoiothermal animals or for their extinction, but these are provisional explanations, and they acknowledge that there could be other explanations. Given that visitors often solve the problems on their own, the organization and interpretation of the provisional reasons for these problems generally involve the use of Lawson's (2010) argument method. Abduction-retroduction and deduction-induction methods, which are used to solve problems resulting from these provisional hypotheses, are shown in Figure 10.

Lawson (2010) and Oh (2012) stated that a claim is a hypothesis suggested to answer a casual question and that it is part of provisional hypotheses generated creatively. However, in this study, the reasons supporting a claim were regarded as proposed provisional explanations because the claim was made as an intermediate theme in the form of a question and because there were multiple reasons to support the claim. Moreover, data as elementary themes support these reasons. These data refer to results predicted by planned experiments and the results of actual experiments. Consequently, these two datasets are compared and are expressed as arrows in both directions in Figure 10, and a conclusion was made based on whether the reason supported the first claim.

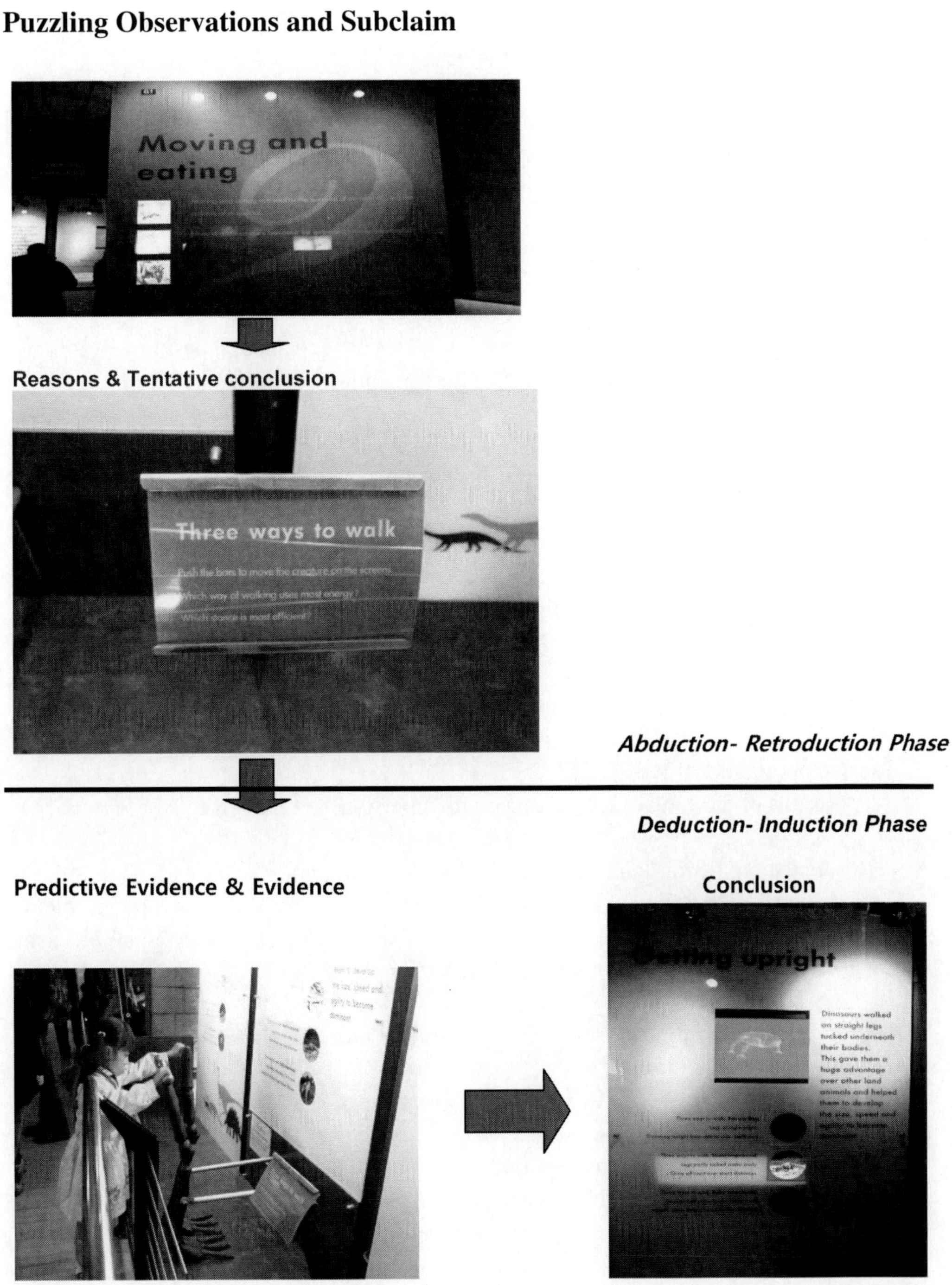

Figure 9. The argument based Deduction -Induction involving Abduction-Retroduction.

■ ***Questions on the argument inquiring into the claim and reasons based on the abduction- retroduction and deduction-induction phases***

Abduction-retroduction phase

1. What caused the puzzling observation?
   Unlike today's reptiles, dinosaurs flourished greatly throughout the earth.
2. What was the most pertinent causal question in this puzzling case?
   How could dinosaurs have flourished so intensely?
3. Have you avoided the "One True Cause" mentality?
   Perhaps, because they walked with straight legs ...
4. What possibility is supported by the evidence? (What explanation best fits everything else we know and believe?) Or which of the possibilities are ranked?
   The best possibility is straight legs ...

Deduction-induction phase

5. Were these tests planned pertinently?
   Legs moved through crawling,
   Slightly improved legs,
   Straight legs improved sufficiently for planned experiments
6. What do the proposed causes lead us to predict regarding future observations? (Deduction).
   Compared to other hypotheses, straight legs are expected to be more efficient in terms of body size, speed, and capability.
7. Tested observational Results?
   Results of the actual experiments on the three types of movements.

8. How do the predictions and new observations? (Induction).
   A conclusion with a proven solution is provided by the following problem-solving argument: "Because they walked with straight legs and were more efficient in all aspects, dinosaurs were able to flourish."

■ ***The Questions regarding the argument inquiring into the claim and reasons based on Abduction and Retroduction***

Abduction and Retroduction phase

1. What caused the puzzling observation?
   Evidence 1, Evidence 2, and Evidence 3... caused the puzzling observation.
2. Pertinent casual question in this puzzling case?
   Were dinosaurs warm or cold-blooded?
3. Have you avoided the "One True Cause" mentality?
   Warm-blood and Cold-blood
4. Which possibility is primarily supported by the evidence? (What explanation best fits everything else we know and believe?) Or, which possibilities are ranked?
   No one knows perhaps dinosaurs were not all the same. Evolving different blood and temperature control systems to match their different ways of life.

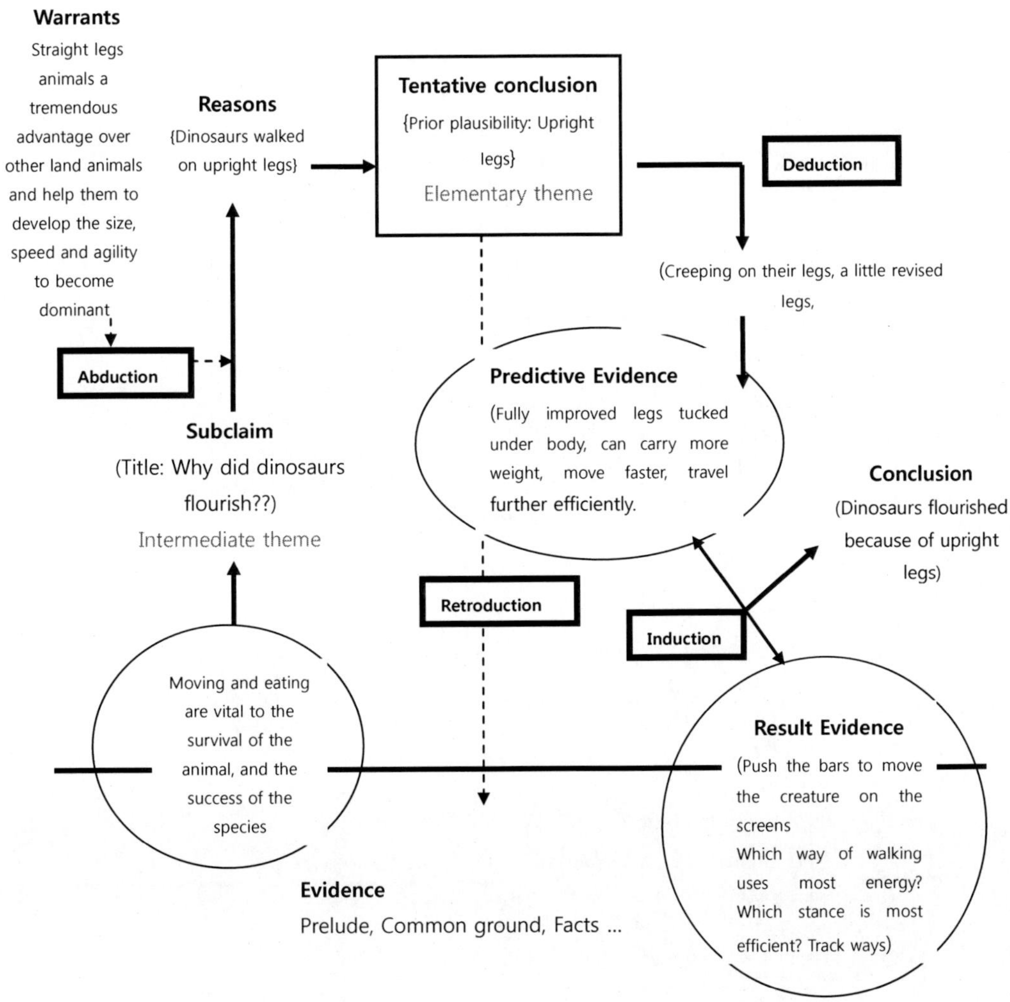

Figure 10. The argument inquiring into the claim and reasons based on the Abduction-Retroduction, and Deduction-induction Phases.

Discussion

First, at the British Natural History Museum, the introduction presented a research problem in the form of a question for visitors to answer. The development (Introduction) consisted of the reasons for the research problem, which were presented as new sub-claims, and supported the exploration of those claims. The conclusion presented solutions for the exposition, and messages for visitors were expressed. The justification of a claim (Booth et al., 2008), which is the most important role of argument, was well organized overall.

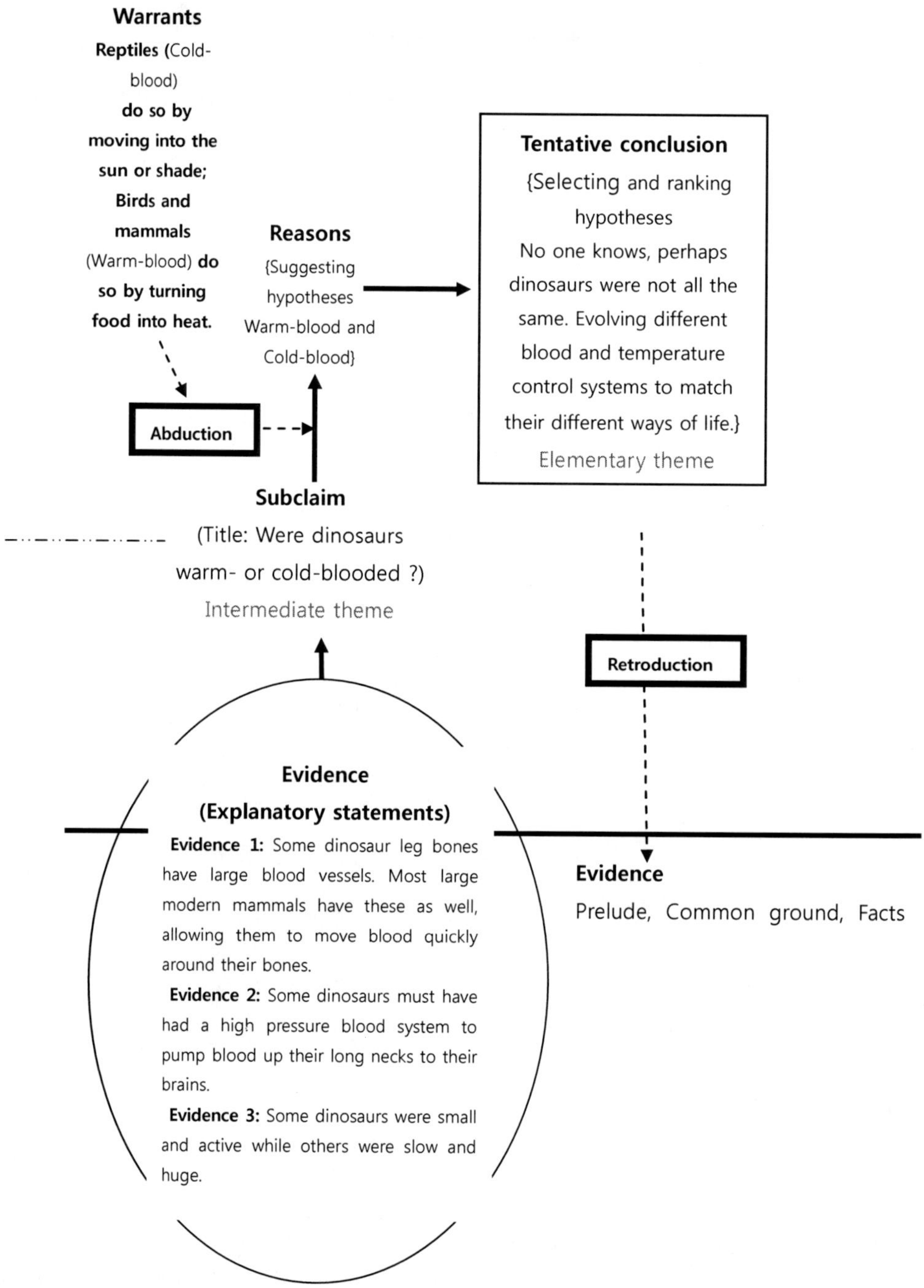

Figure 11. The argument inquiring into the claim and reasons based on the Abduction (Retroduction Phase)

Second, during the microscopic portion, a hands-on argument method (Lawson, 2010) was used to justify a claim (Booth et al., 2008), which is the most important role of argument, and to explore the proposed explanations (i.e., reasons in our study).

Third, the types of hands-on exploration activities included the abduction-retroduction type, which emphasized only the abduction strategy, and the type that considered both the abduction-retroduction and the deduction-induction phases, which considered both the abduction area and the process of proving the hypotheses.

CONCLUSION AND SUGGESTIONS

According to Macdonald (2006, p. 507), it also might be facilitative with regard to another of the science museum's major challengers in the twenty-first century: the composition of its audience:

> Science museums continue to attract great numbers, but the vast majority of those visitors are under the age of sixteen. Since the mid-twentieth century, science museums have largely abandoned the adult audience, and as a result much of the science presented in these places is aimed at a ten-year-old. Educating children about science, convincing them, as these exhibits often try to do, that science is fun, is certainly a worthy enterprise. However, by falling to reach adults, science museums have, however inadvertently, only widened the gap between our scientific literacy and the political choices we are asked to make. Only when the science museum engages a broad adult audience will it help move science into the public realm, and only then can we all approach political decisions as better-informed participants. If science museums take up this challenge, then they will become fully and truly civic instructions. Perhaps by embracing the uncertainty of science and showing it to us, science museums can help us all navigate through an uncertain world (Macdonald, 2006, p.507).

The following are the conclusions and suggestions for the issues affecting not only children but also adults addressed in this study.

First, in the macroscopic portion of this study at the British Natural History Museum, the introduction presented a research problem to be solved in the future in the form of a question. The development consisted of the reasons for the research problem, which were presented as new sub-claims, and supported the exploration of those claims. The conclusion presented solutions to the exposition, and messages for visitors were expressed.

Second, scientists provide provisional explanations or acknowledge the possibility of various explanations. Given that visitors solved the problems on their own, the organization and interpretation of the provisional reasons for these problems generally involved the use of Lawson's (2010) argument method. The Abduction-Retroduction and Deduction-Induction phases used to solve problems resulting from such provisional hypotheses are illustrated.

Third, in the macroscopic portion of the study at the British Natural History Museum, the introduction presented a research problem in the form of a question for visitors to solve. The development consisted of the reasons for the research problem, which were presented as new sub-claims, and supported the exploration of those claims. The conclusion presented solutions for the exposition, and messages for visitors were expressed. The justification of a claim (Booth et al., 2008), which is the most important role of argument, was well organized overall. In the microscopic portion, a hands-on argument method (Lawson, 2010) was used to justify a claim (Booth et al., 2008), which is the most important role of the argument, and to explore the proposed explanations (i.e., reasons in our study).

Fourth, from the perspective that learning science means learning about the culture of scientists, 'authentic science' is evident when students participate in activities with interest and pleasure. Traditionally, science education in schools has focused primarily on convergent activities that lead to a correct answer. The process of continuous exploration is the true form of science. However, students are typically unable to perform scientific activities in which the correct answer is unknown or unavailable or can be explained using various methods. Often, activities do not allow students to express their opinions because the answers are already known. If exploration activities in science museums were evaluated in terms of this problem, students would be more likely to have the chance to associate hands-on activities with thinking activities regarding scientific concepts.

The British Museum of Natural History can be the discovery and learning science museum that integrates constructivism to require active learning situation and thus provide learners with opportunities to operate, explore, or test ideas.

Because of the exhibition form of the latest and largest Gwacheon National Science Museum in Korea, it can be the discovery and learning science museum to integrate the behaviorist view of education, which does not consider the hands-on forms used in the dinosaur exhibition hall in the British Museum of Natural History. Subsequent studies should compare the museum with and study it in relation to the present Gwacheon National Science Museum. Thus, the dinosaur exhibition hall in the British Museum of Natural History was selected.

ACKNOWLEDGMENTS

This work was supported by the National Research Foundation of Korea Grant funded by the Korean Government (NRF-2014S1A5B6037734)

REFERENCES

Afonso, A. S., & Gilbert, J. K. (2007). Educational value of different types of exhibits in an interactive science and technology center. *Science Education,* 91(6), 967 – 987.

Allen, S. (2004). Designs for learning: studying science museum exhibits that do more than entertain. *Science Education,* 88(1), 17 – 33.

Booth, W. C., Colomb, G. G. and Williams, J. M., 2008, *The Craft of Research*(3rd ed.), The University of Chicago Press, Chicago, USA.

Conn, S. (2011). *Science Museums and Culture Wars. In S. Macdonald* (Ed.), A Companion to Museum Studies. UK: Blackwell Publishing Ltd.

Flash, P.A., Kakas, A.C. (2000). Abductive and Inductive Reasoning: Background and Issues. In Flash, P., Kakas, A. (eds.), *Abduction and Induction: Essays on their Relation and Integration* (pp. 1–27). Dordrecht: Kluwer Academic Publishers.

French, L. (2004). Science as the center of a coherent, integrated early childhood curriculum. *Early Childhood Research Quarterly,* 19(1), 150 – 158.

Gelman, R., & Brenneman, K. (2004). Science learning pathways for young children. *Early Childhood Research Quarterly,* 19, 138 – 149.

Hein, G. E. (1998). *Learning in the museum.* London: Routledge.

Hodson, D. 1996, Practical work in school science: exploring some directions for change. *International journal of Science Education,* 18(7), 755-760.

Lawson, A. E. (2010). Basic Inferences of Scientific Reasoning, Argumentation, and Discovery. *Science Education,* 94, 336-364.

Lawson, A. E. (2003). The nature and development of hypothetico-predictive argumentation with implications for science teaching. *International journal of Science Education,* 25(11), 1387-1408.

Lee. S.-K., Lee, S.-K. Kim, C.-J., and Kim, H.-B. (2005). Scenario and Argumentation in Informal Science Learning Material: The Dinosaur Exhibition in Natural History Museum in London, *Journal of the Korean Association for Research in Science Education,* 25(7), 849-866.

Macdonald, S. (2006*). A Companion to Museum Studies.* MA: Blackwell Publishing Ltd.

Magnani, L. (2001). *Abduction, Reason, and Science. Processes of Discovery and Explanation.* New York: Kluwer Academic/Plenum Publishers.

Magnani, L. (2009). *Abductive Cognition: The Epistemological and Eco-Cognitive Dimensions of Hypothetical Reasoning*. Berlin/Heidelberg: Springer.

Oh, J.-Y. (2012). Understanding Scientific Inference in the Natural Sciences Based on Abductive Inference Strategies. In L. Magnani, and P. Li (Eds.), *Philosophy and Cognitive Science: Western & Eastern Studies, SAPERE 2* (pp. 221-237), Heidelberg New York Dordrecht London: Springer-Verlag.

Oh, J.-Y. (2014). Understanding Natural Science based on Abductive Inference: Continental Drift. *The Foundations of Science, 19(2),* 153-174.

Sandifer, C. (2003). Technological novelty and open-endedness: Two characteristics of interactive exhibits that contribute to the holding of visitor attention in a science museum. *Journal of Research in Science Teaching,* 40(2), 121 – 137.

Van Schijndel, T.J.P., Franse, R.K., & Raijmakers, M.E.J. (2010). The Exploratory Behavior Scale: Assessing Young Visitors' Hands-On Behavior in Science Museums*, Science Education,* 94, 794-809.

Wandersee, J. H., Mintzes, J. J. and Noak, J. D., (1994). Research on alternative conceptions in science. In D. L. Gabel (Ed.), *Handbook of research on science teaching and learning,* Macmillan, New York, 598.

Watson, R. 2000, The role of Practical work, In M, Monk & J. Osborne (ed.)., *Good practice in science teaching: What research has to say?*, Open University Press, 57-71pp.

White, R. T. (1996). The link between the laboratory and learning. International Journal of Science Education, 18(7), 761-774.

Williams, J. M. and Colomb, G. G. (2007). *The Craft of Argument,* (3rd ed.), Pearson Education, Inc. New york,

Williams, J. M. and Colomb, G. G. (2003). *The Craft of Argument,* (2nd ed.), Addison Wesley Longman, Inc., New York,

In: New Developments in Science Education Research
Editor: Nathan L. Yates
ISBN: 978-1-63463-914-9

Chapter 8

Dialogical Action Research into Pre-Service Science Teacher Collaboration with a Local High School: A Cultural-Historical Activity - Theoretical Perspective

Seyoung Hwang[1], Namsoo Kim[2], Sang-Hak Jeon[1], Hyun-Pyo Shim[1] and Kum-Bok Ryu[1]
[1]Department of Biology Education, Seoul National University, Korea
[2]Asian institute for Energy, Environment and Sustainability, Seoul National University, Korea

Abstract

In this chapter, we describe the reflexive enquiry process in which a team of researchers has engaged as the teacher educators of the pre-service teachers who participated in a university/school collaborative program. This study focuses on a case, which can inform ways of developing intervention strategies for pre-service teacher development. The program has been experimental as a complementary opportunity to the teaching practicum, undertaken under the support of the partnership between the department and the participating school, and sponsored by the local Office of Education.

A dozen pre-service science teachers participated in the overall process of planning, implementing and reflecting during ten sessions of the inquiry-based biology class. The primary role of the research team has been to provide, not only guidance on instructional content and methods, but also opportunities to 'reflect on action' by engaging in critical dialogue with the pre-service teachers throughout the whole process. In this study, cultural historical activity theory (CHAT) was adopted as a lens through which to identify the nature of learning as reflexively constituted through the interactions and tensions within and between two activity systems, namely research activity (the researchers being the subject) and program activity (the pre-service teachers being the subject).

Over three years, our goal for developing a dialogical mode of intervention strategies has been formed around two questions: i) How can we define and implement an effective co-reflection strategy for enhancing the quality of collaborative teaching? and ii) How can we define and implement inquiry-based science instruction by using laboratory work? In terms of pre-service science teacher professional development, co-reflection strategies have become more fruitful for the PSTs to identify and share issues related to science instruction and to improve accordingly.

Concerning the role of teacher educators, more dialogue is necessary in order to facilitate PSTs to think towards the fundamental objectives of science instruction beyond the management of the program. This case study shows how to become more reflexive teacher educators through gaining insight into its own culture and that of the PST community, and ascertaining a third space for communication and collaboration.

AIM AND CONTEXT OF THE STUDY

In this paper, we describe the reflexive enquiry process in which a team of researchers has engaged as the teacher educators of the pre-service teachers who participated in a university/school collaborative program. In Korea, four years' undergraduate study at a university is the primary route to becoming a qualified teacher at primary or secondary school level. Within the four years of study, however, actual training takes place only during a month-long teaching practicum during the fourth year. Unlike Professional Development Schools (PDS) in the USA, or other similar initiatives (Abdal-Haqq, 1998), there has been little institutional and systematic support for pre-service teachers' professional development from a long-term perspective. In recent studies, many efforts have been made to address such a gap between theory and practice in pre-service teachers' knowledge, by considering effective interventional means, such as mentoring or peer group reflection during a teaching practicum (Oh et al., 2008; Kim et al., 2010). However, little attention has been paid to more systematic ways of pre-service teacher development, which focus on collective inquiry into the cultural learning process by pre-service teachers and researchers within a community.

This study focuses on a case, which can inform ways of developing intervention strategies for pre-service teacher development, by highlighting the experiences and perspectives of teacher educators.

The authors of this paper have been working together on the program in which a group of undergraduate students at the Department of Biology Education at a large, research-oriented university in Korea volunteered as teachers of high school students in a one-year after-school science program. A dozen pre-service science teachers participated in the overall process of planning, implementing and reflecting during ten sessions of the inquiry-based biology class.

The primary role of the research team has been to provide, not only guidance on instructional content and methods, but also opportunities to 'reflect on action' by engaging in critical dialogue with the pre-service teachers throughout the whole process. Therefore, the situation provides an apt context for critical examination of the nascent discourse of intervention constituted through cycles of action and reflection, and the second order analysis of the research process itself, namely, that the study focuses on the formative and reflexive learning process of planning, implementing and reflecting, which a team of teacher educators experienced during their intervention/inquiries over three years.

THE RESEARCH PERSPECTIVE AND METHOD: CULTURAL HISTORICAL ACTIVITY THEORY (CHAT) AND DIALOGICAL ACTION RESEARCH

Over decades, literature on pre-service teachers' learning to teach, and the recognition of the personal and practical nature of teacher knowledge, has been growing within the field of teacher education research (Carter, 1990; Calderhead & Gate, 1993). Science education literature, however, still tended to prioritize the cognitive dimension to the teacher learning process, such as pedagogical content knowledge. More recently, a few researchers began to address the socio-cultural processes of pre-service teacher development from the perspective of enculturation or identity development (Roth & Tobin, 2002; Eick & Dias, 2005; Luehmann, 2007). In this study, CHAT was adopted as a lens through which to identify the nature of learning as being reflexively constituted through the interactions and tensions within and between the two activity systems two activity systems – research activity (the researchers being the subject) and program activity (the pre-service teachers being the subject). In our view, CHAT can contribute to the ongoing discussion surrounding the role of teacher educators and the socio-cultural influences on pre-service teacher learning by illuminating the co-constructive meaning-formulation regarding potentially good science instruction and intervention strategies, as shaped and transformed within multiple activity systems (Engeström, 1987; Edwards, 2008). Figure 1 summarizes the ways in which the research team and the pre-service teachers, as a learning community, interacted with each other.

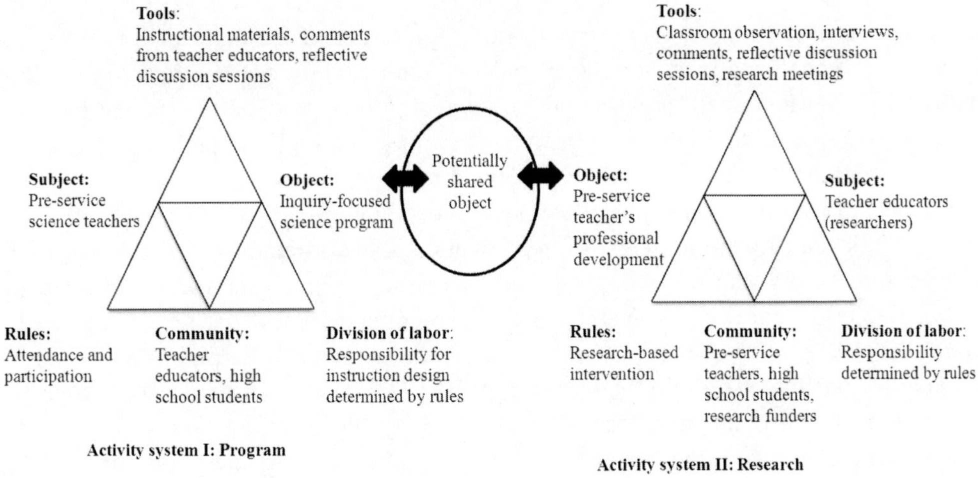

Figure 1. Two activity systems.

In our project, the means of intervention was not pre-determined by the researchers but was pursued in order to be developed in collaboration with pre-service teachers. Importantly, during the intervention/research process, 'co-generative dialogue' (Eldon & Levin, 1991; Roth & Tobin, 2001) was actively pursued as the methodological tool in order to explicate and share the dilemmas, reflections and decisions that we individually and collectively experienced. The dialogical action research method was a useful framework by which to develop our means to co-generative dialogue from the CHAT perspectives because cycles of interaction between action (program activities) and research could be consciously captured and analyzed (Mauer & Githens, 2010). We were interested in linking the theory and practice of pre-service teacher development by combining the first-order action research by developing intervention methods through theoretically-driven analyses of pre-service teachers' teaching, and the second-order action research by revealing the assumptions and beliefs behind our research practice itself. The empirical data sets then included 1) program data, such as classroom observation notes and audio tapes of post-class reflective discussion sessions, and 2) the research team's dialogue and research diaries wherein reflective thinking about our research activity took place. The current paper focuses especially on the insights gained by us concerning the dialectic construction of intervention and inquiry (Elliot, 1993; Orland-Barak & Becher, 2011).

DEFINING THE RESEARCH OBJECT: 'DIALOGICAL' INQUIRY/INTERVENTION

Before the research team formed in 2011, the field program had already been running for three years as a collaborative project between a local high school and the university. During this pre-research period, the role of the university was confined to providing laboratory work resources whereas preparing inquiry instructions was the responsibility of the pre-service science teachers themselves. Participation in the project over one year resulted in a credit for the Educational Service unit, which every education degree student has to earn in order to graduate. Such link to accreditation became an incentive for some PSTs to participate in the program, but for some others, their participation was sustained for more than one year as members of 'Protocol' – a kind of college students' club within the Department of Biology Education, which was formed in 2005 with an interest in biology experiments and has been the core of the program throughout its history. Therefore, the PSTs' experience of teaching biology to high school students could be perceived not only as an individual student's field experience but also as an emerging form of a learning community.

The most distinct feature in the PSTs' participation in the program was found in terms of their rules and the division of labor in the activity system, which themselves showed the culture of reflection and decision-making within the community. At the beginning of each program, the PSTs formed a team in ranging from two-four members who would be responsible for one instruction. In the planning phase, they chose a curricular topic based on past program materials or any new ideas related to biology, and developed it into a three-hour session in which laboratory work was the core activity. During the class, one of the team members would be the lead teacher, and other members and the remaining PSTs sat with the high school students. In many cases, the PSTs and students formed one-to-one relationships in a small group of four PSTs as mentors and four students as mentees. Besides such an explicit arrangement, PSTs' teaching has had certain rules and learning culture with somewhat loose or implicit characteristics, which would be investigated thoroughly during the whole research period. Therefore, when the research team was formed to enhance the quality of the program, it was an encountering of two discrete communities whose agendas and interests did not quite converge naturally.

The initial motivation for conducting research and providing interventions came from a graduate student – H.P. Shim - who had been a core member of the Protocol since its inception and was now studying for a science education degree. He was concerned about how to improve PSTs' participation in the program and the quality of teaching. From his perspective, since not all PSTs shared commitment to the program, not only in terms of the quality of each instruction, which varied greatly but also the fact that a year-long program was not planned systematically. Shim and his supervisor, Prof. S.H. Jeon, given that their main research area was developmental genetics, wanted to collaborate with a science education expert, specifically in the area of teacher education.

They discovered that there was a small amount funding scheme within the science education department aimed at collaborative research among different specialized areas. This was how the team was formed with Dr. S. Y. Hwang, a science education expert, as the lead investigator. During 2011-2012, the team's work focused on identifying the aim of research through intensive literature review works – namely, the object in terms of activity system. Ideas considered at the time included implementing instructional strategies to improve inquiry-based science teaching, investigating PSTs' development of pedagogical content knowledge and measuring students' attainment through pre- and post- surveys, which were based on rather traditional research methods.

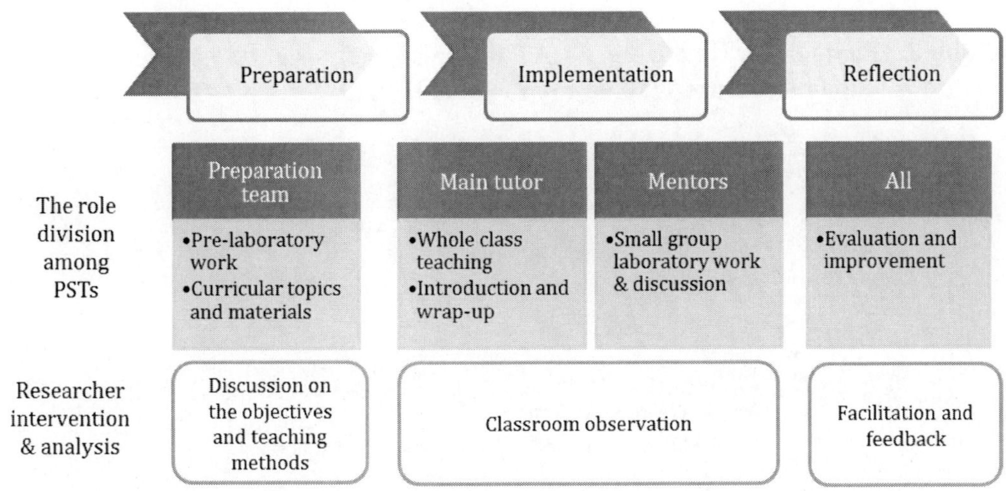

Figure 2. The program arrangement and role division.

On the other hand, the team became interested in the distinct cultural characteristics of the PST community since we realized that we should initially know about the motivations and perceptions of PSTs before considering intervention strategies. Then we began to consider action research as the research method by recognizing the relationship between the research team and the PST community in terms of collaboration, rather than one-way training. Encountering the literature on CHAT gave a fruitful chance for the team to see the ways in which the practice of teaching was shaped through the operation of culturally- and historically-formed objects, norms and artifacts. In 2012, the team succeeded in obtaining research funding for a two-year project, entitled "Dialogical Action Research into a Science Teacher Education Program based on Cultural Historical Activity Theory". The team members included Dr. N.S. Kim who used to be a science teacher, and now a teacher educator, and K.B. Ryu, another graduate student in Prof. Jeon's laboratory. Figure 2 shows the arrangement of program and roles among the PSTs and researchers at this early phase of the inquiry.

In order to define the orientation of our research inquiries and intervention strategies, we conducted a critical review into the current PSTs' professional development strategies in the Korean context (Hwang & Kim, 2014). To summarize the study, we identified four major intervention strategies by characterizing the elements of CHAT and the relationships between them: namely, individualized reflection strategy, expert mentoring, co-reflection strategy, and collaboration between university and school. The analysis highlighted the extent to which each strategy pursued changes in the traditional teaching practicum or dilemmas identified through pursuing the changes in terms of each element of CHAT, which indicates that 'expansive learning' occurs in this process and "produces culturally new patterns of activity" (Engeström, 2001, p.139). The review study implies that in spite of the noticeable difference, a more collaborative and reflective mode of pre-service teacher development was constrained in each strategy due to the dilemmas arising when pursing changes: namely, pursuing changes in one element did not bring about necessary changes in another related element in the system. Based on the review study, we developed a series of questions that can be used to conduct a CHAT-informed analysis of PSTs' professional development in a more 'dialogical' mode for the current research project (Table 1).

Table 1. Questions for conducting CHAT-informed analysis of PSTs' professional development (Adapted from Kim & Hwang, 2013, p.180)

CHAT elements	Questions for characterizing PSTs' professional development
Object	• What is the goal of PSTs' development activity? • Are there any concrete objectives? • How are the objects defined by the participants?
Subject	• Who are the individuals and communities that pursue the goals of the activity? • Who are the subjects? • Who has ownership of the activity?
Community	• Are the membership and members' roles closed or open? • Is there potential for the community to expand?
Division of labor	• How is the division of labor determined? • Is there potential for the roles to be negotiated or changed?
Artifacts and rules	• Who makes the artifacts and rules? • Is there potential for the rules to be negotiated? • Is the negotiation recognized by the members as an important part of the activity?
Outcome(s)	• To what extent can the outcome(s) be shared by the participants? • Does/do the outcome(s) contribute to problem-solving or the formation of a new discourse?
Relationship with other activity systems	• How firm is the boundary between the activity systems? • Is the relationship between the activity systems hierarchical or horizontal?

The questions in Table 1 were further utilized to identify the critical issues related to the ways in which the research team and the PSTs could work together as a learning community. Table 2 shows the initial understanding gained during the first year study about the way in which the program was arranged and the explicit and implicit rules were formed within the community. From the teacher educator's perspective, it seemed important to see these cultural arrangement and norms as sources for further intervention in ways, which facilitate the dialogical mode of interaction. Since then, many dilemmas have been identified, tackled, and are still ongoing, as our understanding about the program/research from the CHAT perspective has increased.

Table 2. Issues of the program identified from the CHAT perspective

CHAT elements	Issues of the program	Data source
Object	The aim of the program was not clearly defined and pursued by PSTs themselves.	Post discussion sessions
	'Inquiry-based instruction' was vaguely defined, and implemented without a coherent framework.	Post discussion sessions Team Seminar
	Also within the research team, different ideas about the role of research prevailed.	Meeting minutes
Subject	A considerable gap exists between the PSTs in the level of passion and interest.	Observation
	Dialogue between PSTs and the research team has progressed slowly, but is yielding some positive effects (see artifacts and rules).	Observation/Meeting minutes/Workshop
Community	PSTs' participation in the program is voluntary, and membership changes each year.	Observation
	Each session is prepared and led by the group of 3-4 PSTs, but others are also obliged to participate in the small group inquiry activities.	Observation
	The research team gives advice during pre- and post-class discussions.	Observation
Division of labor	PSTs' self-confidence on content knowledge and experience influences their actual role in helping students' learning.	Post-discussion sessions
	Therefore, the research team's advice during the pre-instruction session has focused on increasing all PSTs' self-confidence.	Pre-discussion sessions/Meeting minutes
Artifacts and rules	Reflective discussion sessions helped PSTs to feel the need to change the rules and reach their consensus.	Observation/Meeting minutes
	Upon PSTs' request, the research team helped to improve instructional plans and student sheets.	Pre-discussion sessions/Meeting minutes
	Now into the third year, the PSTs feel it natural to make their rules and seek for advice from the research team.	Pre-discussion sessions/Meeting minutes
Outcome(s)	The research team's analysis of PSTs' perception of the purpose of inquiry-based science instruction was shared with PSTs.	Team Seminar

Table 2. (Continued)

CHAT elements	Issues of the program	Data source
	PSTs are more actively participating in the overall process.	Observation
	Three action research projects are under way in a more collaborative mode between PSTs and the research team.	Class reports/Meeting minutes
	Effectiveness/students' learning progress	Student interviews/work sheets
Relationship with other activity systems	Initially, PSTs tended to regard the research team's role as 'observer and assessor', and also within the research team, our role was not clearly defined.	Meeting minutes/Observation
	Now, PSTs feel it natural to ask for help and get feedback.	Meeting minutes/Observation/Pre-discussion sessions

From the initial understanding, our goal for developing a dialogical mode of intervention strategies has been formed around two questions:

1. How can we define and implement effective co-reflection strategy for enhancing the quality of collaborative teaching?
2. How can we define and implement inquiry-based science instruction by using laboratory work?

In the section below, we present the ways in which two questions arose and were then addressed as the research/program activity system evolved interactively. In addressing each question, we specifically use the research reports we produced as the mediating artifact, which informed the educator's empirical understanding about PSTs' perception, and then later guided inquiry/dialogue strategies.

Analysis: Developing Interventions/Inquiries and Making Changes in the Activity System

How Can We Define and Implement an Effective Co-Reflection Strategy for Enhancing the Quality of Collaborative Teaching?

In the literature on PST education, co-reflection has been emphasized as an effective tool for PSTs to identify and critically examine their own thoughts and assumptions on the ideas about good teaching and being a good teacher (Fund, 2000; Lane et al., 2014). In our study/program, reflective discussion was run right after the classroom teaching. During the initial phase of research, the practice of reflective discussion has not yet been established as the 'rule' for every PST's mindset, and from the research team's point of view, the lack of motivation for actively participating in the reflective discussion was part of the major problems in making the program more systematically prepared. Indeed, the PSTs themselves considered reflective discussion as an opportunity for evaluating what went right and/or wrong during the instruction, and what has to be implemented in order for the next session to be well-prepared.

With this awareness, the study conducted during the 2011 program focused on the features eminent in the PSTs' reflective discussion sessions in order to ascertain the teacher educator's role for effective co-reflection.

In the study, PSTs' perceptions on science teaching, based on practical work, were framed in terms of the fit between the sequences of a unit of preparation-actual teaching-evaluation: that is, what the instruction is for (aims) when planning, what a teacher does or is equipped with during the instruction (teacher knowledge), and what students eventually learn (students' response) (Shim et al., 2013).

The data analysis based on these components revealed anything lacking as well as identifying the prevalent features in PSTs' discussion on their teaching practice: in summary, PSTs' discussions were largely focused on the conceptual understanding of science and science process skills rather than inquiry and interest in science (aim and student response), and their teaching methods rather than curriculum knowledge and knowledge of learners (teacher knowledge). Such a discrepancy between the components of each category in PSTs' discourse could be interpreted as the lack of integrative thinking regarding the criteria for judging what constitutes good science instruction. Furthermore, very often, the discourse about 'success' and 'failure' of each instruction was easily formed when the majority of the PSTs were aware of whether something went well or wrong during the laboratory work. Actually, PSTs were very concerned about getting the laboratory work done without any unexpected event happening, rather than perceiving the failure as an opportunity for asking 'why', which could be the relevant pedagogical moment to promote students' participation into inquiry.

Based on our understanding about the PSTs' discourse, formed during the reflective discussions, the role of researchers was actively sought. In fact, it was the researcher (Shim) who acted as the facilitator during the reflective discussions at the time. He himself was not trained to mediate PSTs' discussions, thus his questions were rather randomly thrown in, in order to make anyone speak. It was understood that the limitation in the PSTs' reflection on their teaching was partly led by the absence of the relevant cues, which might otherwise prompt PSTs' reflective thinking better. Therefore, it was naturally the next step to develop the method of questioning, based on the research study (Shim et al., 2013).

In terms of the activity system, this has meant an effort to introduce a new tool for mediating reflective discussion more effectively, based on the mutual understanding regarding the purpose of the reflective discussion for supporting PSTs' more active and natural participation in it. Table 3 summarizes our common awareness on the situation of the PSTs' reflective discussion at that time and how we dealt with it during the reflective discussions the following year.

As the quality of reflective discussion improved, it began to induce many issues to come to the surface, especially in terms of PSTs' consciousness of their own commitment and instructional preparation, such as selecting subject matters. As in Table 4, through discussion, there have been practical changes in the community's rules in order to enhance the quality of the program.

Table 3. Intervention based on the analysis of reflective discussion

Dominant features of reflective discussion in 2011	Intervention strategies for prompting reflection used in 2012
The purpose of instruction was not clear and ideas were not well integrated	• Began the discussion by asking the preparation team to clarify the purpose of instruction • Discussed explicitly 'inquiry'-related themes • Sharpened and reframed the point of discussion towards concrete and integrated idea
No framework or rule for reflection	• Induced as many PSTs' diverse opinions as possible • Used the framework by research paper in order to facilitate integrated thinking regarding instruction • Gave comments on the specific scenes rather than asking overall questions

Table 4. Impact on the instructional design and practice

	2011	2012	2013
Commitment to instruction preparation (subject, division of labor, community, rule)	PSTs teamed up to prepare and lead one instruction in turn, but there were no devices to ensure it.	Along with the strengthened leadership of the leader, rules regarding preparation were actively discussed.	Timeline for preparation was strictly decided upon, and inquiry elements were planned to be included across the program. Data of the past 'reflective discussion' began to be reviewed and improvement was sought.
Selecting subject matters (subject & tool)	Mostly re-using and slightly modifying the existing materials	PSTs began to discuss the potential subject matters	Some PSTs took the initiative for developing new subject matters
Introducing 'mentor's role' (tool, rule, division of labor)	The need for 'mentor's role' – instruction for mentors during group activity was discussed due to the unequal level of preparation among mentors	Actually made and used prototypes	Formulated the content and made it a rule
Reformulating membership rules (community, division of labor, rule)	The problem of free rider(s) was raised due to the uncommitted members whose motivation was mainly earning credits on Educational Service.	Rules on attendance and preparation team became stricter. Mentor – mentee pairs were fixed along with steady attendance. During the main lecture, mentors were to step back rather than engage with mentees.	Stricter rules to earn the credits on Educational Service applied.

New rules were suggested and applied within the PST community over three years, and it can be conceptualized in terms of 'expansive learning' as the tensions and dilemmas between components of the activity system were identified, and efforts made to resolve them. For example, the 'mentor's role' refers to the instructional note for mentors to use when guiding small group activities. This new artifact was a device by which to reduce the gap in mentor's instructional ability, sometimes revealed due to some PSTs' absence from the preparation laboratory work session. By referring to the 1-2 page-long sheets prepared before the actual instruction, PSTs were able to know what questions to ask the students and how to respond to the students' questions, but the downside of using such materials was also obvious: some PSTs relied heavily on it without making conscious efforts to improve their content knowledge and skills for quality interaction. In effect, the rule changes in the PST community through their increasingly conscious 'reflective discussions' tended to be more those of practical strategies for managing the community more efficiently than of fundamental approaches to improving the quality of the program and PSTs' own professional competence. In spite of such remaining issues, the changes in themselves show the co-evolution of the activity system of the PST community, the program and the research.

How Can We Define and Implement Inquiry-Based Science Instruction by Using Laboratory Work?

As addressed in the data of the 2011 reflective discussion, 'inquiry' was not the dominant terminology in PSTs' conceptions about biology laboratory instruction, but as 'inquiry' was explicitly used by the researcher, as when asking, 'What kind of 'inquiry' components were intended during this class?', PSTs gradually used the term, but with various or obscure meanings. Our focus group interview data, classroom observation, and reflective discussions have led us to identify five sustaining characteristics in the PSTs' consciousness of inquiry during the three year period, as follows.

Firstly, as other studies also revealed (Windschitl, 2004; Gyllenpalm et al., 2010), scientific inquiry vs. inquiry teaching were two different concepts: namely, scientific inquiry requires one to explore natural phenomena or questions led by his/her own curiosity whereas inquiry teaching concerns guiding students toward arriving at scientifically correct concepts. This dualistic framework hindered PSTs' thinking beyond currently available laboratory work designs.

Secondly, inquiry referred to more hands-on skills rather than scientific reasoning and the epistemic practice in which scientific theory is proposed and proved. PSTs' focus was to increase students' participation in the laboratory work process itself but without necessarily allowing scientific reasoning during the hands-on activity.

Thirdly, following the second point, it was assumed that the 'thinking' part could be achieved through students' answering the 'open questions' prepared for the instruction.

Fourthly, inquiry was absent during laboratory work.

Fifthly, if inquiry-oriented components of instruction were not effective during the actual class, PSTs tended to return to more traditional teaching methods rather than trying to improve instructional designs or teaching skills. Therefore, even though PSTs became more conscious of inquiry-based teaching in terms of their discourse, their actual teaching practice has become more focused on students' participation in hands-on activities (see Table 5).

Table 5. PSTs' discourse about inquiry-based biology instruction and related actions undertaken in the class

	2011	2012	2013
Discourse	Not very conscious of inquiry	The term 'inquiry' frequently used; theories about inquiry-based instruction were mentioned; emphasis on process skill and scientific reasoning	More initiatives for developing inquiry-based instruction by the PSTs themselves, in spite of ill-definition of 'inquiry'
Instructional practice	Focused on scientific concepts and successful laboratory work	Introduced 'open questions' to induce students' active thinking; developed "What a scientist does" program aiming to enable students' scientific inquiry process	Decided to develop three programs specifically focusing on 'inquiry'; revised "What a scientist does" program with more specific purpose and procedure; emphasis on students' driven activities

This discourse-practice gap was also revealed in the ways in which PSTs interacted with students, as our research paper (Ryu et al., 2014) showed. For example, PSTs' preparation of 'open questions' enabled them to be ready to respond actively to students' answers and discussions, whereas their ability to cope with students' immediate questions depended heavily upon the individual PST's ability or experience. Furthermore, when students asked questions, the mode of interaction between the PSTs and the students tended to be traditional, i.e. confirming and transmitting knowledge, rather than being dialogical, i.e. facilitating scientific reasoning and constructing scientific knowledge together. Also, the division between the two phases of inquiry as laboratory work and discussion was assumed so strictly that PSTs did not allow students' active thinking during the laboratory work. This reveals the PSTs' perception of laboratory work as following the 'cooking recipes', rather than as the process of collecting evidence on scientific theories. In fact, during the preparatory laboratory work sessions, after which the preparation team confirmed the instructional plan, PSTs did not discuss what students would be curious about with regard to the laboratory work or what would be the purpose of the laboratory work for learning scientific concepts. Instead, they focused on practicing the procedure of the laboratory work, which could then be implemented punctually during the actual instruction. Therefore, when the lab work went wrong, the PSTs considered the whole instruction as having gone wrong, without trying to take advantage of students' active thinking about 'why' or 'what happened?'.

Conclusion: Becoming Reflexive Teacher Educators Through Inquiry/Intervention

The whole research process has been a learning experience on the part of teacher educators in (re)-positioning reflexive identities. As we addressed two research reports, research could be conceived as both the result of the observation and analysis, and of the mediating artifact to facilitate mutual reflection. The initially set purpose of research has gone through many revisions and reorientations as our interpretation of 'data' became more focused on how to communicate better with PSTs, and truly support their reflective practice. Practical ideas and strategies of intervention were considered and applied by re-interpreting the data in ways which tackled the assumptions behind the rules and division of labor being taken for granted within the PST community. Through this experience, the role of researcher has been transformed from the outside assessor in the eyes of PSTs, to more of an attentive observer who was willing to help.

Our current concern is how to act more like a co-planner of the program in such a way that our expertise and PSTs' competence can improve at the same time. In terms of pre-service science teacher professional development, co-reflection strategies have become more fruitful for the PSTs to identify and share issues related to science instruction and to improve accordingly. Concerning the role of teacher educators, more dialogue is necessary in order to facilitate PSTs to think towards the fundamental objectives of science instruction beyond the management of the program. Concretely, it is necessary to problematize the enduring discourse around 'inquiry' and to develop prototypes of inquiry-based science instruction with PSTs.

To conclude, the current study tried to contribute to ways of facilitating PSTs' professional development by highlighting the reflexive nature of intervention strategy development from the teacher educators' experiences, especially the ways in which goals and methods of intervention were continuously re-shaped. In doing so, the CHAT-informed analysis was an effective bridge, which enabled us to identify the current situations and then to move towards the necessary changes. The setting of the current study has a distinct feature in terms of the relationship between two activity systems – PSTs and researchers - in that our relationship was rather loose and without institutional obligations. This might imply that if dialogical action research methods became an institutionalized device for PST professional development, more conflicts could be made visible due to the gap in authority between two activity systems. Nevertheless, this case study shows how to become more reflexive teacher educators through gaining insights into its own culture and that of the PST community, and discovering a third arena for communication and collaboration.

REFERENCES

Abdal-Haqq, I.(1998). *Professional development schools: weighting the evidence.* Thousand Oaks, CA: Corwin Press.

Britzman, D. P. (1991). *Practice makes practice: A critical study of learning to teach.* New York: SUNY Press.

Calderhead, J., & Gate, P. (1993). *Conceptualizing reflection in teacher development.* London: Falmer Press.

Carter, K. (1990). Teachers, knowledge and learning to teach. In M. Petetz, R. Bromme, & R. Halkes (eds.). *Handbook of research on teacher education.* NY: Macmillan.

Edwards, A. (2008). Activity theory and small-scale interventions in schools. *Journal of Educational Change*, 9, 375-378.

Eick, C., & Dias, M. (2005). Building the authority of experience in communities of practice: The development of preservice teachers' practical knowledge through coteaching in inquiry classrooms. *Science Education*, 89(3), 470-491.

Eldon, M., & Levin, M. (1991). Cogenerative learning: Bringing participation into action research. In W. F. Whyte (Ed.), *Participatory action research* (pp. 127-142). Newbury Park, CA: Sage.

Elliot, J. (1993). *Reconstructing teacher education*. London, England: Falmer Press.

Engeström, Y. (1987). Learning by expanding. An activity-theoretical approach to developmental research. Helsinki: Orienta-Konsultit.

Engeström, Y. (2001). Expansive learning at work: toward an activity theoretical reconceptualization. *Journal of Education and Work*, 14:1, 133-156.

Fund, Z. (2010) Effects of communities of reflecting peers on student-teacher development – including in-depth case studies, *Teachers and Teaching: theory and practice*, 16:6, 679-701.

Gyllenpalm, J., Wickman, P.,& Holmgren, S. (2010). Teachers' Language on Scientific Inquiry: Methods of teaching or methods of inquiry?, *International Journal of Science Education,* 32:9, 1151-1172.

Hwang, S., & Kim, N. (2014). Analyzing practicum activity from the perspective of cultural historical activity theory. *The Journal of Korean Teacher Education*, 31:1, 141-170.

Kim, HJ., Hong, HG., & Jeon, HY. (2010). A study on the improvement of teaching competence of pre-service science teachers based on the teaching evaluation and reflective journal writings on science class. *Journal for Korean Association of Science Education,* 30:6, 836-849.

Lane, R., McMaster, H., Adnum, J., & Cavanagh, M. (2014). Quality reflective practice in teacher education: a journey towards shared understanding, *Reflective Practice: International and Multidisciplinary Perspectives*, 15:4, 481-494.

Luehmann, A. L. (2007). Identity development as a lens to science teacher preparation. *Science Education*, 91:5, 822-839.

Magnusson, S., Krajcik, J., & Borko, H. (1999). Nature, sources and development of pedagogical content knowledge. In J. Gess-Newsome & N.G. Lederman (Eds.), *Examining pedagogical content knowledge* (pp. 95–132). Dordrecht, The Netherlands: Kluwer.

Maurer, M. & Githens, R. P. (2010). Toward a re-framing of action research for human resource and organization development: Moving beyond problem solving and toward dialogue. *Action Research Journal,* 8:3, 267-292.

Oh, P.S., Lee, S.K., Lee, G.H., Kim, C.J., & Kim, H.B. (2008). Narrative inquiry on students' teachers' teaching experiences with extracurricular science classes of high school: The types and characteristics of the knowledge generated by preservice science teachers. *Journal for Korean Association of Science Education*, 28:6, 546-564.

Orland-Barak, L. & Becher, A. (2011). Cycles of action through systems of activity: Examining an action research model through the lens of activity theory. *Mind, Culture, and Activity*, 18, 115-128.

Roth, W. M., & Tobin, K. (2002). *At the elbow of another: Learning to teach by coteaching*. New York: Peter Lang.

Roth, W-M. & Tobin, K. (2001). The implications of coteaching/cogenerative dialogue for teacher evaluation: Learning from multiple perspectives of everyday practice. *Journal of Personnel Evaluation in Education*, 15, 7-29.

Ryu, K., Shim, H., Lee, E., Jeon, S., & Hwang, S. (2014). Characteristics of Interaction between Pre-service Teachers and High School Students during Small Group Practical Work. *Biology Education*, 42:2, 143-159.

Shim, H., Ryu, K., Lee, E., Jeon, S., & Hwang, S. (2013). Features in pre-service teacher' reflective discussion on their practical work-based teaching. *Journal of Korea Association for Science Education*, 33:5, 911-931.

Windschitl, M. (2004). Folk Theories of ''Inquiry:'' How Preservice Teachers Reproduce the Discourse and Practices of an Atheoretical Scientific Method. *Journal of Research in Science Education,* 41: 5, 481–512.

In: New Developments in Science Education Research
Editor: Nathan L. Yates
ISBN: 978-1-63463-914-9

Chapter 9

DYNAMIC VISUALIZATIONS IN THE CLASSROOM: THE CASE OF PHASE CHANGE

Dalit Levy*

Kibbutzim College of Education, Technology, and the Arts,
Tel Aviv, Israel

ABSTRACT

The chapter presents a study aimed at exploring the advantages of dynamic visualization technology for the development of better understanding of molecular processes. As part of the TELS project (Technology Enhanced Learning in Science), a learning module has been designed in which high school chemistry students conducted virtual experiments with dynamic molecular visualizations of solid, liquid and gas. They interacted with the visualizations and carried out inquiry activities to make and refine connections between observable phenomena and atomic level processes related with phase change. The explanations that students entered in response to pre/post assessments items have been analyzed using a novel scale for measuring the level of molecular reasoning. Results indicated that from pre-test to post-test, students made progress in their level of molecular reasoning, and were better able to connect intermolecular forces and phase change in their explanations. The chapter presents the results through the lens of three improvement patterns: the *extend pattern*, in which more students were able to use an advanced molecular reasoning after interacting with the dynamic molecular models; the *converge pattern*, in which fewer students gave explanations without referring to the molecular point of view after interacting with the dynamic models; and the *shift pattern* combining the previous two, to reach an even higher level of molecular reasoning. The chapter analyses these patterns using the 'ladder of molecular reasoning' metaphor, and discusses how this adds to our understanding of the benefits of interacting with dynamic molecular visualizations.

* Dalit Levy: Kibbutzim College of Education, Technology, and the Arts. E-mail: dalit.levy@smkb.ac.il.

1. INTRODUCTION

Chemistry is often referred to as the "molecular" science, in the sense that understanding chemistry requires an understanding of molecular properties and processes. As such, students of chemistry are required to imagine scientific phenomena on an atomic-scale level and to employ a molecular point of view when explaining these phenomena. Typically, textbooks and classroom instruction include static visualizations and models that help students "see" the symmetries and structural factors involved in chemistry. Students are expected to make connections between these static visualizations and the observable phenomena, as well as to understand that the static representation is just one specific frame of the ever-changing dynamic molecular world. However, research suggests that students have difficulties both in connecting the observable and the molecular points of view (Smith, Wiser, Anderson, and Krajcik, 2006) and in realizing that molecules are always in motion (Pallant and Tinker, 2004). In their report focusing on a learning environment designed to support understanding of chemical concepts, Michalchik et al. (2008) also noted that the understanding of molecular properties and processes has always been a challenge, in a large part because molecules and their properties are not available to direct perception.

Even if molecular phenomena could be clearly demonstrated, research in science education informs us that observation alone might not be enough for learners to develop their scientific understanding. Crouch, Fagen, Callan, and Mazur (2004) compared students learning concepts of physics from different modes of classroom demonstrations and found that those students who passively observed demonstrations understood the underlying concepts no better than those students who did not see the demonstration at all. In his Computing Education Blog, Guzdial (2011) cites Eric Mazur in concluding that in some cases observing a demo is even *worse* than having no demo at all. However, Mazur and his team also showed that students who were actively engaged with a demonstration by predicting the outcome before it had been conducted were better able to recall and explain the scenario posed by that demonstration (Mazur, 2011). A similar idea is reflected by the predict-observe-explain pattern (Kali and Linn, 2008; Linn and Eylon, 2006), according to which students benefit from recording their predictions prior to the demonstration, to be better able to discuss the contradictions after observing it (Wouters, Paas, and Van Merriënboer, 2008).

Today's technology can provide even more ways for students to engage actively with scientific concepts and ideas. In their recent book, Linn and Eylon (2011) assert that "Visualizations that take advantage of advances in technology can enable learners to explore phenomena that are too small (molecules), fast (electrons), abstract (forces), or massive (the solar system) to observe directly" (p. 186). Furthermore, in typical middle or high school classroom settings, where it is impossible to perform real experiments, computer simulations are considered an attractive low-cost alternative (Feurzeig and Roberts, 1999). In describing the simulations designed as part of the CPU project[1], Goldberg (2001) also states that these can help students in collecting both phenomenological and model-based conceptual evidence, as well as in providing multiple representations of such kinds of scientific phenomena. Recent powerful visualization environments such as Molecular Workbench (MW)[2] (Xie and Tinker, 2006; Tinker and Xie, 2008; Xie and Pallant, 2011), NetLogo (Wilensky and Rand, in press;

[1] Constructing Physics Understanding in a Computer Supported Learning Environment. http://cpuproject.sdsu.edu.
[2] By the Concord Consortium. See http://mw.concord.org.

Sengupta and Wilensky, 2009), Connected Chemistry (Stieff, 2005), and PhET (Wieman, Adams, and Perkins, 2008), not only dynamically demonstrate such phenomena, but also enable learners to *interact* with the otherwise inaccessible world. Learning and understanding can be enhanced further by embedding such dynamic visualizations in a learning environment, such as the University of California – Berkeley's Web-based Inquiry Science Environment (WISE), in which students can make their own thinking visible and their own knowledge-integration processes assessable (Kali and Linn, 2008).

Nowadays, many computer simulations can be found on the web as Open Educational Resources (OER) and can therefore, be introduced as an alternative to real experiments in science classrooms at all levels (Panoff, 2009; Finkelstein et al., 2005). Simulation-aided teaching is regarded as an important instructional technology "as it adapts to today's students who grew up in an increasingly digital world and are more accustomed to visual learning" (Xie and Pallant, 2011, p. 122). Dynamic simulations afford learners new ways of visualizing complex domains that may lead to more effective learning (Ainsworth and Van Labeke, 2004; Gerjets et al., 2010) while raising fewer misconceptions (Vosniadou, 2010). They provide visualizations of phenomena and processes that are impossible to see in the real world, yet the experience of which will provide understanding that is it difficult to achieve without such representation (Ainsworth, 2008).

In the case of molecular science, visualization technology enables learners to dynamically model scientific phenomena on the atomic- and even the nano-scale levels (Russell, Kozma, Becker, and Susskind, 2000; Stieff, 2005; Shipley and Moher, 2008; Xie and Lee, 2012). For example, high school students explorations of and interactions with atomic-scale dynamic models of solids, liquids and gas, can lead to a better understanding of the connections between atomic-scale events, such as the overcoming of an intermolecular force, and those events that they can observe at the macroscopic scale, such as a cube of ice melting (Berenfeld and Tinker, 2001; Buckley et al., 2004; Tinker, Berenfeld, and Tinker, 1999; 2000). Other concepts, such as temperature, pressure, equilibrium, and many more can only be understood in terms of dynamic processes (Xie and Pallant, 2011). At the same time, technologies that allow for manipulations of dynamic media, such as deconstructing molecular processes using 2D/3D visualizations (Rundgren and Tibell, 2010), scrutinizing visual data, pacing and segmentation of events (Wouters, Paas, and Van Merriënboer, 2008), reflecting on portions of dynamic media, and generating animations (Bailey and Worth, 2004), can support higher-level thinking and reasoning about molecular processes.

The study reported here was conducted as part of the NSF funded TELS center[3] and involved Molecular Workbench (MW) dynamic visualizations. The development of the MW learning environment began ten years ago with a molecular dynamics engine that calculated in real time the interactions among and between molecules (Fredette, 2012). The software offers educational tools to visualize the collective motions of atoms and molecules based on estimations of classical dynamics and applicable forces. Each run of MW calculates Newtonian approximations of inter-molecular forces to decide how particles will move (Xie and Tinker, 2006). MW has been expanded and currently provides several additional engines for dynamic visualization of DNA transcription and translation, electron clouds, chemical reactions, photonics, biotech and more.

[3] TELS – Technology Enhanced Learning in Science, an NSF supported center headed by Prof. Marcia Linn (UC Berkeley) and Dr. Robert Tinker (The Concord Consortium). See http://www.telscenter.org/.

These computational engines facilitate the instructional design of interactive model-based learning activities for middle school to college science classes, as is stated on the MW website: "The power of MW lies in a set of underlying computational engines that support experimentation at multiple scales and across disciplines"[4].

As the title of this chapter indicates, the focus of this case study is on learning about phases of matter and phase change as part of the high school level Chemistry curriculum.

The rationale for the study derived from the assumption that by manipulating MW dynamic visualizations, by collaboratively interacting with the molecular models, and by reflecting on the interactions, students have the potential to develop a deeper conceptual understanding of the underlying chemical phenomena (Kozma, 2003; Xie and Pallant, 2011). Such an understanding is hereby termed “Molecular Reasoning” (MR). MR is defined as the ability to describe and explain scientific phenomena by referring to the atomic, molecular world (Levy and Tinker, 2008). In other words, MR is the ability to think, reason, and talk about familiar phenomena in a new scientific way (Scott, 1998) using terminology of the atomic scale.

The aim of this case study has been therefore to explore the advantages of dynamic molecular visualizations for offering 'access to new conversation' (Sutton, 1996) about phases of matter, and for the development of a better understanding of the molecular processes of phase change. *The research question* was: *How do students improve in their ability to use molecular reasoning (MR) when explaining phase change, following the interaction with dynamic molecular visualizations*?

In order to answer this question, an online learning module was developed, focusing on "Phases of Matter and Phase Change", while employing dynamic MW visualization tools as part of the online learning process. The implementation of the module in eight public high schools was accompanied with extensive data collection, including online pre/post assessments designed to stimulate thinking about the molecular world. Students' responses to these assessments were analyzed using a scale for measuring the MR level before and after learning with the dynamic MW visualizations. The next sections briefly describe the online learning module, detail the methodological considerations, present the main results and answer the abovementioned research question. The last section draws conclusions and presents some implications from these findings for curriculum developers, teachers, and researchers.

2. The Phase Change Case Study

2.1. The Learning Module

The "Phases of Matter and Phase Change" module is one of more than twenty free online learning units that engage students in scientific inquiry through challenging collaborative activities that emphasize visualization, simulation, and investigation, using information technology (Clark, Varma, McElhaney, and Chiu, 2008; Slotta and Linn, 2009). The module was designed and implemented in high school chemistry classes as part of the TELS online free curriculum (TELS stands for Technology Enhanced Learning in Science, see for example

[4] From Molecular Workbench (MW) website http://mw.concord.org/modeler/moremw.html.

Lee, Linn, Varma, and Liu, 2010). The instructional design followed an inquiry-based, technology-enhanced learning approach (Linn, Davis, and Bell, 2004) and the assumption that the educational value of models emerges from deep interaction with them (Tinker and Xie, 2008). Like other TELS modules, the "Phases of Matter and Phase Change" module was created and implemented using WISE (Linn, Clark, and Slotta, 2003; Liu and Linn, 2011), and integrated novel visualization technologies from the Concord Consortium (Clark, Varma, McElhaney, and Chiu, 2008; Levy, 2009). The learning module utilizes Molecular Workbench (MW) dynamic molecular models of solid, liquid, and gas. Students interact with these visualizations by changing variables, such as heat or pressure (Levy, 2013).

The online learning module is constructed from a set of five web-based learning activities divided into steps. The goal is to develop the ability of learners to employ a higher level of molecular reasoning (MR) when explaining daily phenomena involving phase changes. This goal is achieved by embedding interactive dynamic MW visualizations of the phases of matter in the online learning environment in such a way that the learners' interactions with these dynamic representations are preceded by prediction making and then followed by reflective note taking. The instructional design pattern of *predict-observe-explain* enables learners to make and refine connections between observable phenomena and atomic level processes related to phase change (Linn and Eylon, 2006) and also supports individual inquiry, peer discussions, and reflective note taking. As is also summarized by Wouters, Paas, and van Merriënboer (2008), studies in science education indicate the importance of actively inciting learners to anticipate the outcomes of animated models, for example, by predicting the model's next step or by answering pre-questions.

By the end of the module, the learners are expected to be able to use commonly accepted vocabulary associated with phases of matter and phase changes, describe the role of intermolecular forces in the three phases of matters and in the cases of solid-to-liquid and liquid-to-gas phase changes, describe how intermolecular forces lead to an associated potential energy and its interplay with kinetic energy, express the equivalence of average kinetic energy and temperature, describe the heating curve of water and discuss accurately the reasons for its structure, distinguish between heat and temperature in the case of the heating curve of water, and explain why there is no change in the temperature when melting and evaporation occur, even though heat is added to the system.

2.2. Participants

Chemistry teachers from eight TELS public high schools in North Carolina, Massachusetts, Virginia, and Wisconsin used the module in their classrooms. The implementation of the module in these classes was accompanied with extensive in-class and online data collection, in order to create a rich documentation of the learning process, to gather a wide collection of learners' ideas, and to track changes in those ideas. Overall, data were collected from six hundred students in grades 9-12 that learned using the "Phases of Matter and Phases Change" module as part of their regular chemistry course. As has been often the case in TELS experimental schools, the students worked through the module in pairs, so that the data they entered online were collected and saved as some sort of a shared knowledge (Levy, 2002) rather than an individual response. As Rowell (2002) highlights, shared technological activity is a social situation in which talk not only mediates, but also in part, constitutes the nature of the action. While the talk may focus on specific features of the subject at hand, it also sustains the social relations among participants. In such interplay between subject matter and social relations, the learners in the pair bring their viewpoints into contact in what is regarded in the literature as 'collaboration for inquiry' (Linn and Eylon, 2011). The interaction among pairs while completing the module's activities has indeed been beneficial for eliciting ideas (Sampson and Clark, 2011) as well as for developing individual Molecular Reasoning (MR) ability; however, neither the researcher nor the teachers could distinguish the level of reasoning of individual students from the collaborative written response of the pairs. Thus, although the data analysis reveals common ideas among these students, and although these ideas helped in the construction of the MR scale, as is detailed in section 2.4, this emergent construct was not used to scale, grade, or evaluate individual knowledge, nor individual students.

2.3. Data Collection

The data used in this study came from online pre-test and post-test pair assessments. The ten assessment items were carefully designed by a team of TELS researchers (Lee, Linn, Varma, and Liu, 2010) in order to stimulate thinking about the molecular world and to describe daily phenomena using molecular terminology. The assessment items were all traditional e-assessment items, either in the form of multiple-choice (MC), constructed response (CR), or as a combination of the two. Half of the items asked students to link and connect ideas and give explanations for their conjectures, consistent with the knowledge integration that guides the design of TELS curriculum materials (Linn and Chiu, 2011). Both pre- and post-assessments were administered during regular class periods, shortly before and immediately after running the "Phases of Matter and Phase Change" module, when each pair of students used one computer to enter its answers online.

The data collected from the online tests included 314 pre/post pairs; one pair of pre/post tests for one pair of students. Each pair of students was given an identifying number (for example, pair #257). The teachers were instructed to have their students take the pre-test on the first day of running the online project and to keep the same student pairing throughout the run, so that the same pairs of students would also take the post-test together. That way, the responses of pair #257 in the pre-test could be compared with the responses of the same pair in the post-test. The post-test was usually administered less than two weeks after the pre-test, shortly after the class had finished all the activities of the online module. The two tests were similar, except for a variation in one combined item asking about a heating curve in the pre-test versus asking exactly the same about a cooling curve in the post-test.

The responses to the MC-only items (without the students explaining why the choice was made) were not helpful in expressing levels of molecular reasoning, and therefore, were excluded from the current analysis. Accordingly, the case study reported in this chapter focused on analyzing the data from four items, so that the overall database included ~1200 paired pre-test/post-test written descriptions (~300 paired responses for each of the four items). By consisting of detailed explanations, as phrased by the students themselves before and after the module, the responses for these four items potentially offer the greatest contribution to answering the current research question - *How do students improve in their ability to use MR, following the interaction with dynamic molecular visualization?*

2.4. The "Molecular Reasoning" Scale

The analysis began by examining the different kinds of explanations students gave in response to Item 7; "What happens to water molecules when a cube of ice is taken out of the freezer and left at room temperature?" in the pre-test. In responses like "the molecules expand and melt" (pair #89), "they spread apart" (#214), and "the water molecules start to excite and move around again" (#297), it was quite obvious that students had thought about molecules and thus, revealed a certain level of molecular reasoning, while many other pre-test responses did not mention the molecular point of view at all (even though the question directly requested a description of molecular behavior). As Vosniadou (2010) writes, summarizing the extensive literature on students' misconceptions, many chemistry students believe, like pair #89, that molecules expand when heated, but although based on a misconception, such a reply does at least mention molecules. However, those responses that did mention molecular behavior reveal different levels of molecular reasoning. Those describing how the water molecules "expand" or "spread apart" when a cube of ice is taken out of the freezer, seem to have a more static image of the molecular structure, compared with those who explained how the molecules start to move faster and faster as the cube of ice melts, as was the case with pair #297. The Molecular Reasoning (MR) scale has emerged naturally from the above observation. Recall that MR has been defined previously as the ability to describe and explain scientific phenomena by referring to the molecular world (Levy and Tinker, 2008). After carefully screening all the written responses to Item 7 in the pre-test, given by the 314 pairs of high school students from various TELS schools, four types of explanation were recognized with regard to MR. Explanations of one type did not refer at all to the molecular world and thus, were categorized as "non MR". The other three types of explanation all mentioned the molecular world. The more common types included phrases and keywords that were interpreted as either expressing a basic static view of the molecular world and thus, categorized as "static MR", or a more advanced dynamic view categorized as "dynamic MR". Very few pre-test responses hinted at an even higher level of reasoning, referring to the intermolecular forces in explaining the phase change phenomena. These rare responses were categorized as "intermolecular MR". Therefore, the four types of MR can be placed on an ordinal scale constructed of the zero level of no MR, and then the static, dynamic, and intermolecular levels of understanding in thinking about phases of matter and phase change (see Table 1 for example responses).

The development of the coding scheme using the MR scale, involved a team of TELS researchers as well as leading members of the MW development team. According to this coding scheme, explanations that do not contain any trace of thinking in the molecular level are scored "0", and those that do are scored "1" (static), "2" (dynamic), or "3" (intermolecular). Interestingly, the scale also reflects the historical development of thinking in the atomic level. The historical point of view can be found in a story about the 1999 Nobel Prize winner in Chemistry, Prof. Zewail (from CalTech). The story goes back to 1944, when Pauling, another CalTech professor, won the Nobel Prize in Chemistry. In the reporter's words: "Pauling worked from crystallographic data, and his bonds were static, stable, and enduring. Now, 45 years later, Zewail has set those bonds in motion, making them as alive and dynamic as chemistry itself"[5].

[5] Retrieved September 15, 2012 from http://www.zewail.caltech.edu/nobel/Zewail_Feature.pdf.

Table 1. The Molecular Reasoning (MR) Scale

Score	Description	Examples[6]
3	Intermolecular MR Dynamic MR + including correct description of the weakening of intermolecular forces/bonds	"The molecules start to move faster and break the bonds holding them together" (#18) "It melts because the molecules gain energy, motion, and they can overcome cohesive forces and come apart" (#44) "The cube of ice absorbs energy from the environment and begins to warm up. This increases the kinetic energy and the molecules move faster and the intermolecular bond between them decreases. This makes the ice change to water" (#225)
2	Dynamic MR Referring to molecular motion (specifically mentioning the molecules moving) -OR- Referring to the role of intermolecular forces in phase change (without specifically speaking about molecular motion, as in response #268)	"The molecules change state because of the differences in heat. The molecules can move around easier in a liquid rather that a solid where they are close together" (#94) "The water molecules start to move faster because of the energy flowing from the surrounding to the ice cube. As the water molecules start to move faster the ice cube starts to melt" (#232) "With the room temperature added, the water molecules start to excite and move around again, so which in turns reverts the ice cube back into water form" (#297) "The ice molecules separate in the freezer because the bond weaken. So when it reaches room temp the bonds come back together again". (#268)
1	Static MR Speaking about molecules, but not about their motion.	"They become scattered because of phase change from solid to liquid" (#16) "The water molecules are tightly packed together when the ice cube is first taken out of the freezer and then they slowly become further apart as the temperature of the ice cube increased until the water molecules reach the liquid state" (#113) "The water molecules would breakdown and would not be compacted together because the ice cube is melting" (#277) "The molecules have more space" (#292)
0	No MR	"The Cube of ice will melt, but will melt slowly depending on the room temperature" (#13) "It begins to melt from the temp. change in the room" (#289) "It's melting because the cube is at room temperature and the cube is solid at a freezing temperature. It loses energy" (#301)

[6] These responses were entered online by real students answering Item 7 in the pre-test: "What happens to water molecules when a cube of ice is taken out of the freezer and left at room temperature?", including the typos.

According to the MR scale, the highest score of molecular reasoning is given to explanations that not only describe the motion of the molecules, but also refer to the role that intermolecular forces play in the phase change related phenomena.

It is regarded as the highest level because without consideration of intermolecular forces, one gets stuck when trying to explain melting, for example.

Even if students correctly describe the molecules moving faster and faster as heat is added, that accelerated motion alone does not explain why the solid suddenly becomes a liquid (Schmidt, Kaufmann, and Treagust, 2009). For a fuller account, one must also describe how that energy of motion is used to weaken the intermolecular forces.

Thus, for example, while using the MR scale to code the explanations students gave the pre-test item dealing with the heating curve, it was quite clear that students were capable of recognizing that the flat parts of the heating curve were related to the phase change (probably because they learned that in the past), but they failed to explain why the temperature stayed constant during these times. Therefore, in the pre-test students could answer a question like "What happens to the temperature during a phase change?" (stays constant), but they could not answer the question "Why does it happen?" (energy of heat is used to weaken intermolecular forces) because they lacked the knowledge about the role that intermolecular forces play in this process. In such cases, even using the 2nd dynamic level of the MR scale, describing the change in molecular motion, cannot lead to a full explanation.

The next analytic step consisted of coding all the responses in the database using the MR scale. The pre-test responses for each of the four assessment items were coded first, independently from the post-test responses. The coding itself consisted of carefully reading each response in the database and scoring it according to the scale. Two assessment experts, as well as one of the TELS chemistry teachers, helped in confirming the trustworthiness of the coding of partial sets of responses.

As a last step, the resulting scores for each pre-test item were matched with the resulting scores for each post-test item, to construct the final dataset of MR scores. Out of 314 pairs of students in all the classes participating in this study, the pre/post MR scores of 241 pairs could be matched and have been included in the final dataset.

3. Results

The MR scale represents an ordinal scale and therefore, the sign test was used to compute the significance of the changes in the MR score from pre- to post-test, and to answer the research question. The results are presented next using the metaphor of 'climbing the ladder of molecular reasoning'.

3.1. Improvement Patterns

Examination of the scores of the four items revealed significant differences in the positive direction between pre-test and post-test responses. Table 2 presents the detailed frequencies.

The box plots in Figure 1 below give a graphical image of the concentration of the scores and also show how far the extreme values are from most of the data (Dean and Illowsky, 2012). The two leftmost box plots show, in the case of Item 5, that the middle 50% of the pre-test scores fall inside the 0–1 range (with only *one* MR level 3 score, 0.4% of all pre-test responses), whereas the range of the middle 50% post-test scores extends up to MR level 2 (with twelve MR level 3 scores, 5% of all post-test responses).

A slightly different improvement pattern is shown in the box plots for Items 6 and 7. In both cases, the middle 50% of the pre-test scores fall inside the 0–2 range (no MR level 3 scores in Item 6; only two in Item 7, 0.8%). In the post-test scores, both ranges converge up into the 1–2 range (with nineteen MR level 3 scores in Item 6, 7.9%; forty-one in Item 7, 17%). The improvement pattern in the case of Item 9 is the most noticeable: for the pre-test, the middle 50% of the scores fall inside the 0–2 range. For the post-test, that range "climbs" and the middle 50% of the scores fall inside the 1–3 range (eighteen MR level 3 scores in the pre-test, 7.5%; ninety MR level 3 scores in the post-test, 37%). In summary, three improvement patterns were found: *extend* (like in the case of Item 5), *converge* (like in the cases of Items 6 and 7), and *shift* (like in the case of Item 9). In the *extend pattern*, more students tend to use dynamic MR terminology (level 2) after interacting with dynamic MW models.

Table 2. Frequencies of differences and p-values

	Item 5	Item 6	Item 7	Item 9
Negative Differences (post<pre)	31	21	29	21
Positive Differences (post>pre)	84	88	103	147
Ties (post=pre)	126	132	109	73
Total (N)	241	241	241	241
Z	-4.849	-6.322	-6.354	-9.644
Significant? (<0.001)	Yes	Yes	Yes	Yes

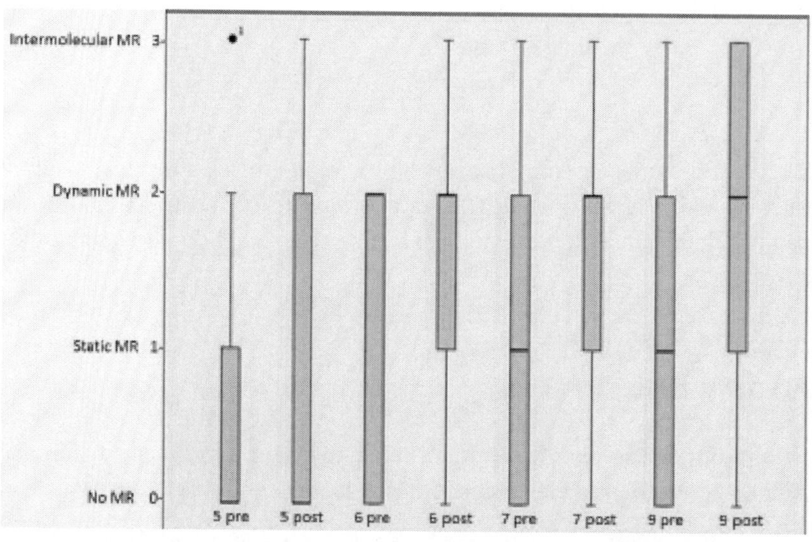

Figure 1. Pre- and Post- MR levels for all 4 assessment items (N = 241 pairs).

In the *converge pattern*, students who interacted with the dynamic MW models tend less to respond in a "no MR" fashion (level 0). In a *shift pattern*, more students tend to use the higher levels of MR (levels 3, 2), and at the same time fewer students tend to respond in a static/no MR fashion (levels 1, 0) after their interactive experience with the dynamic MW models.

3.2. Climbing the Ladder of Molecular Reasoning: Pretest Results

In all of the above-mentioned improvement patterns, students clearly climbed above the static level of molecular reasoning, employing a more dynamic point of view. This section analyses further the students' starting point.

From the pre-test columns of the final dataset it is evident that prior to using the TELS "Phases of Matter and Phase Change" module, most pairs of students were either thinking about the molecular world in static terms only, or were not thinking about molecules at all and therefore, their responses scored 0. The percentage of "non-MR" responses in the pre-test was the highest for Item 5 (68%); whilst about one-third of the responses were categorized as "non-MR" for the other items (see Figure 2).

The combined percentage of the two *low MR levels* in the pre-test was again the highest for Item 5 (85%), ranging from 64% for Item 9 to 45% for Item 6 (see Table 3).

All four assessment items directed the students to include some kind of molecular terminology in their answers. Item 6 even deliberately directed the students to describe molecular motion - MR level 2 - in their explanations:

Item 6. If you want something to dissolve fast, you should mix it with:

(a) Hot water. (b) Cold water.

Explain why, referring to molecular motion.

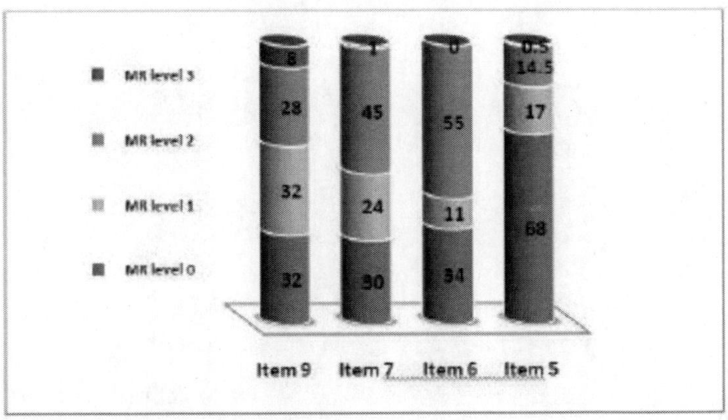

Figure 2. Percentages of responses in the pre-test.

Table 3. Percentages of lower versus higher MR levels pre-test responses

MR level	Item 9	Item 7	Item 6	Item 5
Low 0+1	64	54	45	85
High 2+3	36	46	55	15

Therefore, it is not surprising that the combined percentage of the two *high MR levels* in the pre-test was the highest for Item 6 (55%), and that a little more than half of the pairs (132 out of 241) were able to give explanations like; "The molecules are moving faster (have a higher avg. kinetic energy) and thus dissolve substances faster due to more rapid particle movement compared with cold water" (pair #20, MR score 2). Nevertheless, a large enough number of pairs (109 out of 241, 45%) constructed low level MR pre-test responses even when instructed to include molecular motion in their answers to Item 6, like pair #276; "The molecules expand quicker[7] in cold than on hot water" (scored 1) or pair #226; "the hotter the water more energy it has and will help to dissolve" (scored 0)[8].

In summary, as in the last two examples, the vast majority of the pre-test responses for the four items did not mention molecular motion at all. As can also be seen in Figure 2, only 0%–8% of the pre-test responses were scored as MR level 3, showing clearly that the construct of 'intermolecular forces' was not a part of the students' conceptual framework prior to learning with the "Phases of Matter and Phase Change" module. As is described next, the percentage jumps in the post-test, and the difference is attributed to the visualization of intermolecular forces offered by the Molecular Workbench models used.

3.3. Climbing the Ladder of Molecular Reasoning: Posttest Results

Figure 3 and Table 4 show that the combined percentage of responses categorized as MR levels 0 and 1 drops, while the combined percentage of responses categorized as MR levels 2 and 3 increases significantly; 100% increase for Items 5 and 9 from pre-test to post-test, and about 50% increase for Items 6 and 7.

Comparing the percentages of the highest MR level 3 in the pre-test (8%, Item 9, Figure 2) and the post-test (37%, Item 9, Figure 3), it is evident that the highest numbers appeared in response to Item 9 in both the pre-test and the post-test.

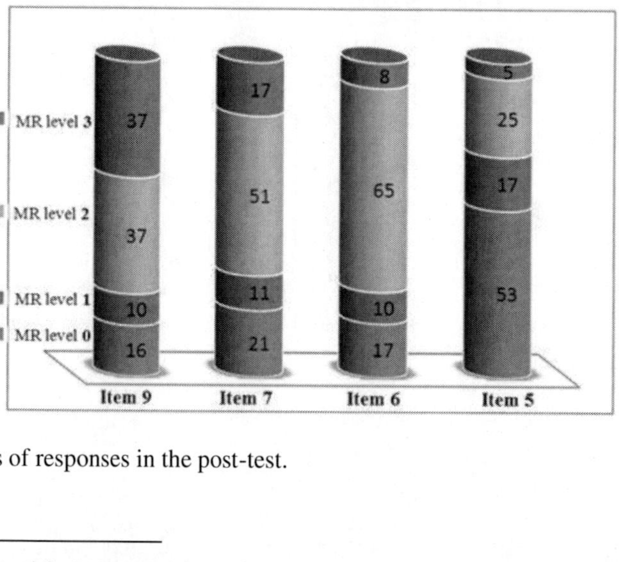

Figure 3. Percentages of responses in the post-test.

[7] The typos appear in the original response.

[8] As the coding scheme does not classify responses as being 'right' or 'wrong', correct answers like the one written by pair #226 could still be scored 0 because of not referring to molecules at all.

Table 4. Percentages of lower versus higher MR levels post-test responses

MR level	Item 9	Item 7	Item 6	Item 5
Low 0+1	26	32	27	70
High 2+3	74	68	73	30

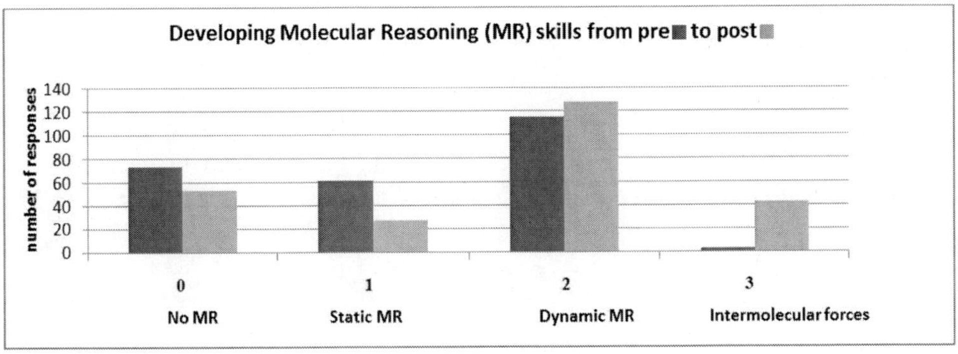

Figure 4. Distribution of MR levels in Item 7 responses.

This result can be attributed to the term "intermolecular forces" being specifically mentioned in Item 9 and not in the other items. However, in spite of the direct instruction, the pre-test responses showed a very limited prior knowledge of that level, while the number of "3"s in the post-test was almost five times higher than in the pre-test. As is shown in Figure 4, the jump is even higher with regard to Item 7 - from two pre-test responses of MR level 3 (0.8%) to forty-one in the post-test (17%).

In addition, Figure 4 shows that much fewer responses were scored lower-level in the post-test (light bars) than in the pre-test (dark bars), hinting at the *converge* improvement pattern, as suggested earlier. In this case, students clearly employed a more dynamic point of view in the post-test. This climb in the ladder of MR is attributed to the interaction with the dynamic Molecular Workbench models throughout the module.

3.4. Summary of Results

The case study reported in this chapter has been initiated by the question: How do students improve in their ability to use molecular reasoning (MR) when explaining phase change, following the interaction with dynamic molecular visualizations? First and foremost, a clear answer offered by the results is that the high school students that participated in this case-study significantly improved in their MR ability and were able to construct explanations of a higher level after learning the "Phases of Matter and Phase Change" and interacting with the dynamic MW models it contains.

Furthermore, the "how" question has been answered using additional lenses. Through the lens of improvement patterns, three patterns have been found: in the extend pattern, more students used dynamic MR after interacting with the dynamic MW models; in the converge pattern, fewer students used "no MR" explanations after interacting with the dynamic MW models; and the shift pattern combined the previous two, to reach an even higher MR level.

Through the lens of climbing the MR ladder, it was found that for the majority of the high school students, molecular motion played only a small part in their conceptual framework, and intermolecular forces were not at all part of that framework prior to interacting with the dynamic MW models. A moderate improvement was found in the ability of these students to include dynamic MR in their explanations after interacting with the dynamic model; a significant progress was found in the ability to consider intermolecular forces in their explanations.

Using an additional qualitative analysis of the responses (Levy, 2013), progress was found for those single pairs that provided MR level 3 responses at the pre-test, when their post-test responses reflected even better understanding of the role of intermolecular forces during processes of phase change – a progress that can be attributed to interacting with dynamic MW visualizations throughout the learning with the "Phases of Matter and Phase Change" module.

CONCLUSION

The behavior of large numbers of molecules each following individual rules of motion, and the related study of heat and phase change, is a complex subject for high school students to understand (Smith, Wiser, Anderson, Krajcik, and Coppola, 2004). The literature suggests some reasons for the difficulty, which include the statistical understanding involved, the abstract nature of thermodynamics as a theoretical system (Wilensky, Hazzard, and Froemke, 1999), the students' inability to directly experience and observe these processes (Meir et al., 2005), and the separation of the micro-level phenomena from the macro-level phenomena (Dori and Hameiri, 2003). This case study highlights an additional source of difficulty, namely the traditional instructional focus on static visualizations of the ever-changing dynamic molecular world.

Textbooks have a static two-dimensional nature and in many cases are insufficient for providing the scaffolding students need to understand phenomena on the molecular level. The results of this case study suggest that students should be given the opportunity to watch and follow what is otherwise too small to be seen and too fast to be traced, as well as to interact with these dynamic visualizations, in order to reflect upon their interactions, and to improve their ability to explain phase-change-related phenomena. Using the scale for measuring the conceptual change in terms of the level of molecular reasoning, the results show that the opportunity is more than a promise, and that students are indeed improving their molecular reasoning skills following such a technology-enhanced learning experience.

One concern that arose from the results is that although significant progress was found in the students' ability to consider intermolecular forces (MR level 3) in post-test explanations, the third level of molecular reasoning remained relatively rare after one week of learning, indicating that many students did not make the most of the web-based dynamic visualizations during this period.

On the other hand, the results also reveal the potential of using such technology for those students who did not have the prior capacity of such an advanced level of thinking about the molecular world. In their comprehensive discussion on how to make visualizations valuable, Linn and Eylon (2011) detail both the controversy, and suggest paths to success in learning with technology-enhanced visualizations in general and with dynamic visualizations in particular. Dynamic molecular models are often multifaceted visualizations of complex scientific phenomena, and as such may cause confusion for learners "who interpret them in unintended ways" (Linn and Eylon, 2011, p. 217).

A key idea of these authors is to take into account successful knowledge integration (KI) patterns when designing dynamic visualizations for science learning. Kali and Linn (2008) add four design principles for better incorporation of visualizations in general and dynamic visualization in particular, into curriculum materials; included amongst them are reducing visual complexity and using multiple linked representations.

The role of dynamic visualizations in science education is clearly becoming increasingly important (Rundgren and Tibell, 2010) and new technologies are assumed to better promote effective learning of complex topics such as phase change. Dynamic visualization technology, such as Molecular Workbench, enables learners, teachers, designers, and researchers to develop and follow a reasoning process based upon interaction with dynamic components that are essential to the understanding of atomic molecular theory and thus, provide a glimpse into better understanding (Staudt, 2008). While many agree that dynamic visualizations are helpful, recent reviews concerning their educational value are inconclusive (Linn and Eylon, 2011). More research needs to be carried out to substantiate the effectiveness of learning with dynamic visualizations (Xie and Pallant, 2011).

The results from such a line of research can provide valuable feedback to educators, instructional designers, and educational researchers alike. While this case study contributes significantly to current knowledge concerning science learning with dynamic visualizations in a high school setting, the implications can be extended to other settings and areas in which the power of dynamic visualizations has been discussed. In particular, we need to orient research to questions such as: How do teachers support learning with dynamic visualizations? What forms of instructional guidance are most supportive of developing dynamic kinds of reasoning? What forms of assessment are appropriate for dynamic constructs?

In summary, although the scientific field addressed here has been specific, the contribution of dynamic and interactive visualizations tools can be extended to other scientific fields and thus, are of importance for a broad community of science educators and instructional designers. Fields other than science might also benefit from considering the ideas presented in this paper for assistance in learning and assessment of those fields.

ACKNOWLEDGMENTS

The author wishes to thank the Concord Consortium developers and researchers and Dr. Galit Ashkenazi-Golan for their invaluable help. This work was partially supported by the NSF under the TELS grant. Any opinions, findings, and conclusions expressed in this chapter are those of the author.

REFERENCES

Ainsworth, S. (2008). The educational value of multiple-representations when learning. In: J. K. Gilbert, M. Reiner and M. Nakhleh (Eds.), *Visualization: Theory and practice in science education*. Dordrecht: Springer, 191-208.

Ainsworth, S. and Van Labeke, N. (2004). Multiple forms of dynamic representation. *Learning and Instruction*, 14, 241-255.

Bailey, E. and Worth, B. (2004). *Making Thinking Visible: Enhancing Media Literacy Instruction*. Unpublished design study report, Stanford University School of Education (Prof. Daniel Schwartz).

Berenfeld, B. and Tinker, R. (2001). Revolutionizing Science Teaching with Atomic Scale Models. *Paper presented at the National Education Computer Conference* (NECC), Chicago, IL.

Buckley, B., Gobert, J., Kindfield, A. C. H., Horwitz, P., Tinker, R., and Gerlits, B. (2004). Model-based teaching and learning with BioLogica: What do they learn? How do they learn? How do we know? *Journal of Science Education and Technology*, 13(1), 23-41.

Clark, D. B. and Sampson, V. (2008). Assessing dialogic argumentation in online environments to relate structure, grounds, and conceptual quality. *Journal of Research in Science Teaching,* 45(3), 293-321.

Clark, D. B., Varma, K., McElhaney, K., and Chiu, J. (2008). Design rationale within TELS projects to support knowledge integration. In: D. Robinson and G. Schraw (Eds.), *Recent Innovations in Educational Technology that Facilitate Student Learning* (pp. 157-193). Charlotte, NC: Information Age Publishing.

Crouch, C., Fagen, A. P., Callan, J. P., and Mazur, E. (2004). Classroom demonstrations: Learning tools or entertainment? *American Journal of Physics,* 72(6), 835-857.

Dean, S. and Illowsky, B. (2012). *Descriptive Statistics: Box Plot*. Retrieved September 15, 2012 from the Connexions Web site: http://cnx.org/content/m16296/1.12/.

Dori, Y. J. and Hameiri, M. (2003). Multidimensional analysis system for quantitative chemistry problems - symbol, macro, micro and process aspects. *Journal of Research in Science Teaching,* 40(3), 278-302.

Feurzeig, W. and Roberts, N. (1999). *Modeling and Simulation in Science and Mathematics Education*. New York: Springer.

Finkelstein, N. D., Adams, W. K., Keller, C. J., Kohl, P. B., Perkins, K. K., and Podolefsky, N. S. (2005). When learning about the real world is better done virtually: A study of substituting computer simulations for laboratory equipment. *Physical Review Special Topics - Physics Education Research*, 1(1), 10103-10110.

Fredette, M. (2012). Curriculum update: Biology evolved. *THE Journal (Transforming Education through Technology)*, 39(6), 16-19.

Gerjets, P., Imhof, B., Kühl, T., Pfeiffer, V., Scheiter, K., and Gemballa, S. (2010). Using static and dynamic visualizations to support the comprehension of complex dynamic phenomena in the natural sciences. In: L. Verschaffel, E. de Corte, T. de Jong, and J. Elen (Eds.), *Use of External Representations in Reasoning and Problem Solving: Analysis and Improvement.* London: Routledge. 153-168.

Goldberg, F. (2001). Some Roles of Computer Technology in Helping Students Learn Physics: Computer Simulations. *Paper presented at International Conference on*

Computer and Information Technology in Physics Education, Manila, Philippines. Retrieved September 30, 2012 from http://www.sci.sdsu.edu/CRMSE/old_site/FG/pub1.pdf.

Guzdial, M. (2011). Eric Mazur's Keynote at ICER 2011: Observing demos hurts learning, and confusion is a sign of understanding. *Computing Education Blog*, Aug 17, 2011. Retrieved September 26, 2011 from http://computinged.wordpress.com/2011/08/17/eric-mazurs-keynote-at-icer-2011-observing-demos-hurts-learning-and-confusion-is-a-sign-of-understanding/.

Kali, Y. and Linn, M. C. (2008). Designing effective visualization for elementary school science. *The Elementary School Journal*, 109(2), 181-198.

Kozma, R. (2003). Material and social affordances of multiple representations for science understanding. *Learning and Instruction*, 13(2), 205-226.

Lee, H.-S., Linn, M. C., Varma, K., and Liu, O. L. (2010). How do technology-enhanced inquiry science units impact classroom learning? *Journal of Research in Science Teaching*, 47(1), 71-90.

Levy, D. (2002). Collaborative conceptual change: the case of recursion. *J. Intell. Syst.* 12 (2):113–135.

Levy, D. (2009). Dynamic assessment tools for dynamic constructs: The case of molecular reasoning. *Paper presented at the National Association of Research in Science and Technology (NARST) 2009*, Garden Grove, CA.

Levy, D. (2013). How Dynamic Visualization Technology can Support Molecular Reasoning. *Journal of Science Education and Technology* 22(5): 702-717.

Levy, D., Tinker, R. (2008). Links between dynamic representations of atomic-scale phenomena and molecular reasoning. In: Eshet-Alkalai, Y., Caspi, A., Geri, N. (eds.) *Proceedings of the Chais conference on instructional technologies research 2008: Learning in the technological era*. The Open University of Israel, Raanana, Israel, pp. 66–70.

Linn, M. C. and Chiu, J. L. (2011). Combining Learning and assessment to improve science education. *Research and Practice in Assessment*, 5, 5-14.

Linn, M. C., Clark, D. and Slotta, J. D. (2003). WISE design for knowledge integration. *Science Education*, 87, 517-538.

Linn, M. C., Davis, E. A. and Bell. P. (2004). Inquiry and technology. M. C. Linn, E. A. Davis and P. Bell (Eds.) Internet Environments for Science Education (pp. 3-28). Mahwah, NJ: Lawrence Erlbaum Associates.

Linn, M. C. and Eylon, B.-S. (2011). *Science Learning and Instruction: Taking Advantage of Technology to Promote Knowledge Integration*. New York, NY: Routledge.

Linn, M. C. and Eylon, B.-S. (2006). Science education: Integrating views of learning and instruction. In: P. A. Alexander and P. H. Winne (Eds.), *Handbook of educational psychology* (2nd ed. pp. 511-544). Mahwah, NJ: Erlbaum.

Liu, J. L. and Linn, M. C. (2011). Knowledge integration and Wise engineering. *Journal of Pre-College Engineering Education Research* 1(1), 1-14.

Mazur, E. (2011). The scientific approach to teaching: Research as a basis for course design. *Keynote presentation at the International Computing Education Research Workshop (ICER 2011), Providence, RI.* Retrieved September 26, 2011 from http://mazur.harvard.edu/sentFiles/MazurTalk_1712.pdf.

Meir, E., Perry, J., Stal, D., Maruka, S., and Klopfer, E. (2005). How effective are simulated molecular-level experiments for teaching diffusion and osmosis? *Cell Biology Education* 4, 235-248.

Michalchik, V., Rosenquist, A., Kozma, R., Kreikemeier, P., Schank, P., and Coppola, B. (2008). Representational resources for constructing shared understandings in the high school chemistry classroom. In: J. Gilbert, M. Reiner and M. Nakhleh (eds.), *Visualization: Theory and practice in science education.* New York: Springer.

Pallant, A. and Tinker, R. F. (2004). Reasoning with atomic-scale molecular dynamic models. *Journal of Science Education and Technology,* 13(1), 51-66.

Panoff, R. (2009). Simulations deepen scientific learning. *ASCD Express,* 4(19). Retrieved Aug 25, 2012 from http://www.ascd.org/ascd_express/vol4/419_panoff.aspx.

Rowell, P. M. (2002). Peer interactions in shared technological activity: A study of participation. *International Journal of Technology and Design Education*, 12, 1-22.

Rundgren, C.-J. A. and Tibell, L. (2010). Critical features of visualizations of transport through the cell membrane: An empirical study of upper secondary and tertiary students' meaning-making of a still image and an animation. *International Journal of Science and Mathematics Education*, 8(2), 223-246.

Russell, J., Kozma, R., Becker, D., and Susskind, T. (2000). *SMV: Chem; Synchronized Multiple Visualizations in Chemistry*. New York: John Wiley.

Sampson, V. and Clark, D. (2011). A comparison of the collaborative scientific argumentation practices in two high and two low performing groups. *Research in Science Education,* 41(1), 63-97.

Schmidt, H.-J., Kaufman, B. and Treagust, D. F. (2009). Students' understanding of boiling points and intermolecular forces. *Chemistry Education Research and Practice,* 4(10), 265-272.

Scott, P. (1998). Teacher talk and meaning making in science classrooms: a Vygotskian analysis and review. *Studies in Science Education*, 32(1), 45-80.

Sengupta, P. and Wilensky, U. (2009). Learning electricity with NIELS: Thinking with electrons and thinking in levels. *International Journal of Computers for Mathematical Learning*, 14(1), 21-50.

Shipley, E. and Moher, T. (2008). Instructional framing for nanoscale self-assembly design in middle school: A pilot study. Paper presented at the *Annual Meeting of the American Educational Research Association*, March 24-28, 2008.

Slotta, J. D. and Linn, M. C. (2009). *WISE Science: Web-based Inquiry in the Classroom.* New York: Teachers College Press.

Smith, C., Wiser, M., Anderson, C. W., Krajcik, J., and Coppola, B. (2004). *Implications of Research on Children's Learning for Assessment: Matter and Atomic Molecular Theory.* Paper commissioned by the Committee on Test Design for K-12 Science Achievement, Center for Education, National Research Council.

Smith, C. L., Wiser, M., Anderson, C. W., and Krajcik, J. (2006). Implications of research on children's learning for standards and assessment: A proposed learning progression for matter and the atomic-molecular theory. *Measurement: Interdisciplinary Research and Perspectives*, 4(1 and 2), 1-98.

Staudt, C. (2008). Alternative assessments with snapshots. *Spring 2008 @Concord Newsletter.* Available online at http://www.concord.org/publications/newsletter/2008-spring/

Stieff, M. (2005). Connected chemistry – A novel modeling environment for the chemistry classroom. *Journal of Chemical Education,* 82(3), 489-493.

Sutton, C. (1996). The scientific model as a form of speech. In: G. Welford, J. Osborne, P. Scott (Eds.), *Research in Science Education in Europe*. London: Falmer Press. 143-152.

Tinker, R., Berenfeld, B. and Tinker, B. (1999). *HOMS: Hands On Molecular Science: Annual report to the National Science Foundation (REC-9980620).*

Tinker, R., Berenfeld, B. and Tinker, B. (2000). *Molecular Workbench: Annual report to the National Science Foundation (REC-9813485).*

Tinker, R. F. and Xie, Q. (2008). Applying computational science to education: The Molecular Workbench paradigm. *Computing in Science and Engineering*, 10(5), 24-27.

Vosniadou, S. (2010). Instructional considerations in the use of external representations: The distinction between perceptually based depictions and pictures that represent conceptual models. In: L. Verschaffel, E. de Corte, T. de Jong, and J. Elen (Eds.). *Use of Representations in Reasoning and Problem Solving: Analysis and Improvement*, London: Routledge, 36-54.

Wieman, C. E., Adams, W. K. and Perkins, K. K. (2008). PhET: Simulations that enhance learning. *Science*, 322(5902), 682-683.

Wilensky, U. and Rand, W. (in press). An Introduction to Agent-based Modeling: Modeling Natural, Social and Engineered Complex Systems with NetLogo. Cambridge, MA: MIT Press.

Wilensky, U., Hazzard, E. and Froemke, R. (1999). GasLab - an extensible modeling toolkit for exploring statistical mechanics. *Paper presented at the Seventh Annual European Logo conference EUROLOGO 99*, Sofia, Bulgaria.

Wouters, P., Paas, F. and Van Merriënboer, J. J. G. (2008). How to optimize learning from animated models: A review of guidelines based on cognitive load. *Review of Educational Research,* 78, 645-675.

Xie, Q. and Lee, H-S. (2012). A visual approach to nanotechnology education. *The International Journal of Engineering Education*, 28(5), 1006-1018.

Xie, Q. and Pallant, A. (2011). The Molecular Workbench software: An innovative dynamic modeling tool for nanoscience education. In: M. S. Khine and I. M. Saleh, (Eds.), *Models and Modeling: Cognitive Tools for Scientific Enquiry*. Springer, 121-132.

Xie, Q. and Tinker, R. F. (2006). Molecular dynamics simulations of chemical reactions for use in education. *Journal of Chemical Education,* 83, 77-83.

In: New Developments in Science Education Research
Editor: Nathan L. Yates
ISBN: 978-1-63463-914-9

Chapter 10

NAVIGATING THE SCIENTIFIC LANDSCAPE VIA A WRITING ACROSS THE CURRICULUM PROGRAM: OUTCOMES, POSSIBILITIES AND CHALLENGES

Ingrid A. M. McLaren*
University of The West Indies, Mona, Jamaica

ABSTRACT

This paper provides insight into issues related to a university-based Writing Across the Curriculum (WAC)/Communication Across the Curriculum (CAC) initiative in the sciences. It focuses specifically on the dialectical implications of cross disciplinary collaboration, the key role played by research findings emanating from this intervention in informing writing praxis in general, and in the sciences in particular, and specific concerns and challenges related to program sustainability. The experiences of other institutions employing a WAC initiative serve as a basis for comparison, support and direction as do the assertions of researchers, experts and practitioners of WAC pedagogy and scientific communication. It is anticipated that the issues highlighted in this paper will serve to inform similar initiatives.

FOREGROUNDING WAC IMPLEMENTATION

Writing in Science: Doing, Knowing and Understanding

Socio cultural constructivists adhere to the notion that learning is intrinsically linked to meaning-making - a semiotic process- that is shaped, transformed reproduced, and transmitted through language. This theoretical framework of which language and learning are significant components, has critical implications for knowledge-making, understanding and

* Ingrid A. M. McLaren, PhD: University of The West Indies, Mona, Jamaica. E-mail: ingridmclaren@gmail.com.

communication in science, and this is evidenced by Halliday and Martin's (1993, para 4) assertion that the scientific system of "technical terms and nominalized grammar" constitute a "language in which theories are constructed". Previous to this, Postman and Weingarter (1971) had posited that since knowledge is inseparable from language, and a discipline is a way of knowing, understanding the language of the discipline is critical to learning.

Moreover, the nature of many scientific –related activities such as the outlining, validating establishing, and reconstructing of theory clearly attest to the need for literacy in science to facilitate the understanding and communication of knowledge. In support Osbourne (2002, p. 206), has contended that the specialized discourse involved in science calls for competence in "talking, reading and writing science". In fact Norris and Phillips (2003, p. 254) argue that rather than being a mere vehicle for the transmission of knowledge, literacy is "constitutive of science itself".

Therefore since language is central to science and science literacy, and also constitutes a way of doing science, constructing understanding and communicating knowledge to others, it follows that no meaningful learning may occur in science without competence in engaging with texts, and expressing ideas. This is entirely consistent with the position of the National Research Council, (NRC), (1996) that text and verbalization are important components of the social construction of scientific knowledge. In a similar vein, The National Science Education Standards (NSES) (1996, p. 21) has referred to science as "a way of knowing characterized by empirical criteria, logical arguments and skeptical review".

Given the perceived value of science literacy to conceptual learning and expression within the classroom, the Writing Across the Curriculum (WAC) movement has been deemed to be the ideal tool for promoting and facilitating this link between literacy education and disciplinary study. Fertile theoretical grounding for this pedagogical approach was laid by Emig`s (1977) constructivist -influenced article "Writing as a Mode of Learning," that posited that writing functions as as a unique and invaluable way for students to understand and integrate information. Later, in 2010, Thassis and Porter referred to WAC as an institutional initiative geared at assisting "teachers across disciplines in using student writing as an instructional tool in their teaching (p. 538)".

The appeal of WAC pedagogy is based on two fundamental principles---Writing to Learn, referred to as the expressivist/cognitive mode of WAC, and Writing in the Disciplines (WID) whose aim is to help students master the discourse of their disciplinary community. WID is guided by social constructivist theories which acknowledge the pivotal role of specialized discourse communities in knowledge making. The primary goals of this approach is to help students write in the mode expected of members of their disciplinary community, and to help students think like professionals by teaching them to express themselves like professionals.

Support for the effectiveness of WAC as a tool in science literacy is evidenced in the contentions of science educators and researchers. For instance, the importance of teaching students the principles of effective writing in the sciences has been attested to by Moore (1993) and Robertson (2004). In fact, Ellis (2004) has found the quality of students' approaches to writing and the way they think about writing as a way of learning, to be closely related to their level of achievement. Similarly, Lerner (2007, p. 192) has highlighted the importance of using writing in science "to inculcate students into the research and communication practices of scientists".

Additionally, science teachers have been encouraged to assign writing activities which enable students to hone their skills in areas such as constructing arguments by using details from direct observations, explaining concepts, and reporting results so effectively that recipients will be able to adequately interpret and evaluate the information, and use it appropriately (Feldman, Anderson and Mangurian, 2001). Such activities (Keys et al., 1999) must be perceived by students as being meaningful and authentic and should match the conceptual structure of a topic and broader curricular goals.

The findings of researchers offer some evidence of the effectiveness of this pedagogical approach to writing and learning in the sciences. For instance, Bocolo and Mason (2001) have shown that writing activities in the sciences can be a useful aid in conceptual comprehension and retention as well as in promoting reflection on such information especially if there is a forum for feedback/discussion. And Hand, Prain and Yore (2001) found that writing activities improved students' ability to communicate more effectively, based on the fact that they consistently scored better on subsequent tests and displayed superior performance on higher-order questions compared to their counterparts in the control group whose writing tasks were not directed.

In addition to this, Cervetti, Pearson, Barber, Hiebert, and Bravo (2007) found that when science and literacy lessons are integrated students display greater skill in both areas. More recently, McLaren and Webber (2009) found that ecology students who were exposed to WAC/WID strategies displayed marked improvement in performance in essay examinations and performed at a significantly higher level than their counterparts who were not thus exposed.

WAC/WID IN CONTEXT

The increasing implementation of WAC/WID strategies generally and in the science classroom across and outside of the United States, specifically, has occurred in the context of growing concern among academics, (Lillis and Turner, 2001; Ganobcsik-Williams, 2006; Byrne, 2007), and science educators (Jerde and Taper, 2004; Moore, 1994), about the writing competence of students. They contend that generally, undergraduate students have not learnt to write effectively in scientific formats and that the majority of scientific writing problems observed are related to the documents' organization, tone, clarity and concision. For this reason these and other researchers (Bangert-Drowns, Hurley and Wilkinson, 2004) have promoted a closer integration of writing with education in all disciplines, not only to improve writing skills but also to facilitate better understanding of subject matter.

These sentiments concerning the dearth of relevant writing skills among students are generally consistent with those of the teaching staff at our institution, particularly those involved in the teaching of science. In fact, the quality of science students' writing has been repeatedly identified as a common weakness in their examination performance on terminal assessments at the secondary and post-secondary level. In support of this, the report of biology examiners of the regional assessment body (Caribbean Examinations Council, 2007a) pointed to "generally vague" responses of the candidates coupled with their apparent inability to write in a logical sequence, even though the students may have some idea of what they intend to express to the examiners of Biology (Caribbean Examinations Council, 2007a).

Consequently, these examiners recommend that students' examination preparation should be focused on "practicing how to interpret and answer questions clearly, concisely and to the point" (Caribbean Examinations Council, 2007a).

Further, lecturers in a number of departments of the Science Faculty found many weaknesses in students' writing at all stages of the degree programme, more so at the entrance level. There was also concern that although these students had been exposed to the principles of academic writing in their Level 1 English Language Foundation courses, there was little evidence of a transfer of these skills to their science courses.

This issue of writing transfers among first year students has been investigated by Herrington (1985), Beaufort (2007), Bergmann and Zepernick (2007), and Wardle (2007). Nearly all findings on writing transfer indicate that if students fail to recognize similar features across writing situations and tasks, they will not be successful in transferring writing skills. Although students often have been taught writing processes and skills that would assist them throughout their educational careers, these studies show that they are often unable to draw upon that knowledge and instead perceive each situation as entirely new and foreign.

In fact Herrington (1985) found in her qualitative examination of the writing in two college chemistry courses, that students failed to see any similarities between the writing tasks and required skills in each course despite the many similarities Herrington perceived between the tasks. She also reported that students saw each course as a unique learning situation instead of adapting their prior knowledge to the new context.

Findings such as these appeared applicable to our context where students themselves reported seeing very little connection between the skills acquired in their English Language Foundation courses and those required of content/disciplinary courses. Furthermore, the fact that lecturers in many of their courses placed more emphasis on students` grasp of course content than on writing competence would probably have justified the perception of many students that writing proficiency was not essential to success in their other courses. This position is consistent with that of Boud (1998, p. 24) who states: "Assessment methods and requirements probably have a greater influence on how and what students learn than any other single factor".

WAC/WID IMPLEMENTATION

Phase 1: Exploring the Landscape, Breaking Ground

In a context such as the one described, the implementation of a WAC program was deemed highly appropriate and was implemented in two phases. Phase 1 of our WAC initiative involved a 2- year writing project that was conceptualized as one which would "integrate English Language teaching and usage into all aspects of the education of students in order to address and rectify the failure of many students to experience writing as an integral component of their university education in general, and their academic discipline in particular" *(The Writing Centre Proposal, 2006).*

Notwithstanding, our goal of undertaking a campus-wide initiative was not realized as funding was provided for the WAC initiative in only one Faculty--the Faulty of Pure and Applied Sciences. This was due to the fact that the request for "one-off" funding for an

English language initiative which eventually led to the WAC Project came from a member of the Faculty of Pure and Applied Sciences who had applied for and received a grant for this endeavour.

The members of the Life Sciences Department showed relatively high and positive levels of response and this was no doubt due in large part to the cooperation and support from the Head who was instrumental in securing an attendance of nearly 80% of his staff members at our training workshops. This was the department which showed the highest levels of interest and engagement amongst the academics and this Head's support of WAC was seen as a response to the concerns of his staff members regarding the poor writing skills of students. WAC strategies included Entrance/Exit Slips, Role Playing, Lensing, Draft-Feedback-Redraft, Journaling.

In contrast to the positive response of the Head of Life Sciences, the Head of Chemistry offered little support for WAC and we thus attempted to 'sell' this initiative to individual staff members in this Department after failed attempts to get them to agree to a convenient time for attending a workshop. Our success rate here was lower than in Life Sciences, as we were only able to encourage three staff members (18%) to formally integrate WAC strategies into their courses.

Although other faculty members in Chemistry and Life Sciences and a few in the Departments of Physics and Mathematics and Computer Sciences also incorporated WAC strategies into their delivery of content, this was not done in a consistent nor structured way.

In order to gain further insight into the reasons for involvement or lack thereof, we interviewed both participants and non participants. In the first instance those lecturers who decided to participate in the implementation of WAC strategies were ones who:

- were deeply concerned about the poor writing skills of their students
- perceived writing as integral to students' personal development and critical thinking skills
- viewed good writing skills as necessary to explaining and clarifying concepts in their course
- were concerned about the poor communication skills on the part of some graduates from their faculty, which they saw as leading to weak performance on the job

Conversely, non participants attributed their reluctance to incorporate WAC based learning activities into the delivery of courses to concerns regarding:

- the additional workload that this would entail;
- the fact that student attendance at tutorials was not compulsory and for this reason it would be difficult to fully engage students and give feedback on writing; and
- the fact that enhanced focus on students' writing skills could not be adequately accommodated in the present system of course evaluation and assessment.

Notwithstanding, this initial Phase of WAC intervention proved to be extremely fruitful in terms of research and publications and produced significant results. Details of these results are revealed and discussed further in this paper.

Additionally, the initial "foray" into the sciences on the part of the English Language Section team paved the way for a fuller and deeper incursion into the sciences as evidenced by Phase 2`s WID/CAC initiative (Writing in the Disciplines / Communication Across the Curriculum) which was geared towards infusing Writing and Speaking into specific courses.

Phase 2: Crossing Disciplinary Boundaries, Inhabiting the Space

Boundaries and Borders carry a certain mystery and fascination;
They imply a transition between realms of experience, and states of being;
They draw a line between life as lived in one place
And life as lived in another.
(Ryden, 1993, p. 1)

The implementation of a Communication across the Curriculum (CAC) project in the Faculty of Pure and Applied (FPAS) Sciences constituted an outgrowth of Phase 1's WAC intervention and was also spearheaded by the author in her capacity as Writing Project Coordinator. The primary goal of this project was to design and implement a (CAC) programme within the (FPAS), which would help students move beyond "general academic writing or novice approximations of disciplinary writing to internalizing the communication-thinking practices of professionals in their field" (Bransford et al., 2001). Beginning with the Departments of Chemistry and Life Sciences whose members had participated in the previous WAC project, the CAC implementation process was designed to involve the selection of four courses from each of these departments; two courses at Level 2 (2nd year) and two at Level 3 (3rd year).

One course at each of these levels for each of these departments would be designated as writing intensive and the other as speech intensive. These courses should be ones which are currently so positioned within the curriculum that most students are required to take them and this would ensure the exposure of most students in these departments to a writing and speaking intensive course at levels 2 and 3. Generally, the benefits of broadening the science curriculum to include reading, writing, speaking and listening skills were seen as critical to enhancing academic achievement, attitude, confidence levels and self-images. Thus, students would be afforded the opportunity of becoming independent, life long learners by active participation in their own learning as opposed to being passive note takers.

Furthermore, this type of infusion was viewed as being aligned to enhanced awareness on the part of educators of the value of communication in all disciplines. Indeed, helping students meet the target competencies of professional practice, teaching them effective teamwork and collaboration and enhancing their ability to understand and argue with visual data are recognized as widespread needs, particularly in the field of science (Shuman et al., 2005).

The nature of the intervention which involved "border-crossing" i.e. cross/inter disciplinary engagement and activities, called for the development of an effective interdisciplinary collaborative model, which would ensure as far as possible, the successful establishment of a CAC programme in the science faculty. The model itself involved the Writing Fellow who delivered instruction on scientific writing to students in chemistry and life sciences courses selected for writing infusion. The same approach obtained for the Speech

Infusion component of the intervention where a Public Speaking Instructor delivered related modules in designated courses (McLaren, 2010).

The formulation of such a model has been reported by researchers such as Emerson et al. (2006) to involve multiple challenges, some of which include an imbalance of power, difficulty in adequately integrating the skills of members of different disciplines and conflicting agendas (McLaren, 2010).

Added to these difficulties would be the socio-cultural dynamics of the institution itself, particularly in the context of compliance and a tradition of non-intra and inter-disciplinary collaboration.

Moreover, the absence of a clear institutional mandate was another critical consideration for the Writing Coordinator (WC) as it was thought that this might impact the forging of meaningful and viable alliances across disciplines i.e., outside the discipline of Humanities to which she belonged. In fact, although the project was supported in word and deed by the Dean of the science faculty, through the allocation of funds, there was no general mandate for the science teaching staff to infuse writing or speech into their courses. This lack of explicit endorsement would possibly involve extra effort on the part of the WC in persuading the teaching staff of the merit of a CAC programme and also in inducing them to infuse writing and speaking into their courses.

Inextricably linked to the laying of this collaborative foundation, were subsequent negotiations related to actual implementation. Such negotiations centered around the selection of courses in which writing was to be integrated, the role of the WC and science staff in the implementation of WAC, the revision of course outcomes and assessment to reflect the integration of writing, and, most importantly, the allocation of time and space in the current syllabus of courses selected for infusion. These latter two considerations proved to be the most challenging and a distinct drawback of a lack of formal institutional endorsement, as in the first case, the revision of course outcomes and assessment had to be delayed as these were subject to administrative approval. In the second case, it was left to the discretion of the lecturers involved to decide how much time would be allocated to the integration process and where exactly in their course schedules such an infusion would occur (McLaren, 2010).

Thus in many cases the pre-conceived notions of the WC, as to what should ideally constitute speaking and writing infusion into courses had to be abandoned. These had been based to a large extent on the status quo concerning the design of and criteria for Writing Intensive (WI) and Speech Intensive (SI) courses, outside of our context, at universities overseas. For this reason, the building of a collaborative framework involved the giving and yielding of ground, with the scientists retaining more of their ground and the Writing Fellow claiming less space than had previously been anticipated(McLaren, 2010).

For instance, the original plan to designate two courses at Level 2 and two at Level 3 in each department (Chemistry and Life Sciences) for writing and speaking infusion at each level had to be modified. In the case of Chemistry, only two courses were selected, one at Level 2 as WI and one at Level 3 as SI, for the Analytical Chemistry Major. The lecturers thought that these courses were the ones which would benefit most from infusion and additionally would capture the majority of students pursuing chemistry courses.

In the case of Life Sciences, the lecturers thought that students in Level 3 courses, the majority of whom were in their final year of study, should be the benefactors of two SI courses and one WI course. Thus, no SI course would be implemented at Level 2 and only one was designated as WI at this level. The rationale was that those at Level 3 who were about to embark on professions would be better prepared at the point of leaving by the infusion of writing and speech into their courses.

Another case in point was the allocation of time to the delivery of speaking modules where a previously planned 21/2 hour allocation per week over 10 weeks was reduced to a mere 1hr over a six week period in the case of Chemistry and seven weeks in the case of Life Sciences.

Moreover these sessions were slotted into breaks in lab sessions in both instances, and thus were often viewed by students as "interruptions". For this reason, the Speaking Instructor had difficulty in achieving the recommended 15 minutes per session to be allocated to "graded oral communication assignments, that might include interpersonal, group, and/oral presentational activities" (http://www.ysu.edu/ger/OralIntensive/Course.doc).

Notwithstanding, our focus on a common goal (that of enhancing students' communicative competence) enabled us to address these issues with flexibility and willingness to compromise in key areas. Furthermore, although, the 'reach' of the initiative was far less extensive than we had projected, as was the case with Phase 1 WAC implementation, we nonetheless viewed our 'border-crossing' as invaluable in terms of the forging of robust interdisciplinary alliances and the successful merging of mutual interests and agendas.

REVIEWING THE LANDSCAPE

Outcomes: Implementation and Research

An objective review of WAC/WID/CAC implementation outcomes at our institution is best situated within an analytical framework such as the ones proposed by Anson (2006) and Condon and Rutz (2012).

In the first instance, Anson has developed a model representing a continuum of engagement that represents individual initiatives and defines the institutional and instructional context where WAC takes place and the different levels of Inquiry. At the same time, Condon and Rutz (2012) have designed a taxonomy which describes the key characteristics of four types of WAC Programs at different stages of implementation, ranging from Foundational to Institutional.

Anson's (2006) model graphically represents individual contexts of activity for WAC initiatives along a continuum (Figure 1) ranging from efforts at the individual level to the formalized institutionalization of WAC programs. This model enabled us to situate Phase 1 of our WAC/WID at the 'Ad Hoc Team Partnership' point of the continuum and Phase 2 at the 'Department' point which involved "a more formalized configuration" (Anson, 2006, p. 102) as obtained at Rice University where trained consultants "work closely with faculty in science and engineering to plan and implement communication instruction in existing courses" (Rice University, n.d.). Thus the model not only indicated some movement forward regarding our implementation efforts but it also offered us a realistic perspective of our position relative to our ultimate goal--that of full Institutional endorsement and support of the WAC/CAC program.

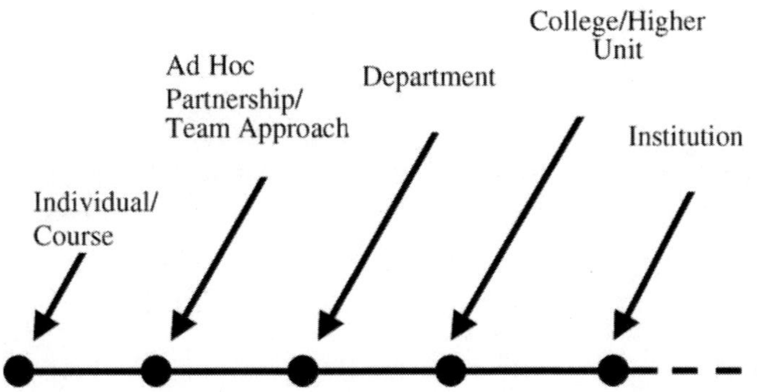

Figure 1. Contexts of Activity -- Anson (2006).

At the same time, based on Condon and Rutz's (2012) Taxonomy of WAC programs which describes features of the 'Foundational', 'Established', 'Integrated' and 'Institutional' stages of WAC Program implementation, we perceived that we were at Stage 1 (Foundational) in terms of funding as our Program was still "largely volunteer effort, sometimes with minor reassigned time Dependent on good will from umbrella (provost, dean, etc.)" (p. 362) compared to the more advanced stages such as the 'Established' where "Program has own budget, though often on temporary funding; Program identity emerges: space, staffing, programming become more visible and regularized" and the stage of 'Institutional/Change' where "Substantial, permanent institutional\funding for well-defined and established roles and personnel" (p. 362), were distinct characteristics.

However, in terms of 'Primary Goals' we were closer to the 'Established stage' as we had moved from the Foundational "Problem-based Statement of purpose" to the continuation of "Faculty development and missionary models", leading others to "serve the WAC agenda, and creating "reliable, continual archives of materials, policies, evolution of program—history" (p. 362).

Thus, the Anson (2006) model and the Condon and Rutz (2012) Taxonomy are invaluable indicators of milestones concerning our efforts to attain full institution –wide WAC endorsement. The initial WAC implementation stage --Phase 1 --lasted approximately 2 years while Phase 2 (CAC) which started in 2010 is still ongoing without expansion to other Departments in the science faculty, or in fact to any of the other Faculties and Departments at the University. The courses which were originally selected for writing and speaking infusion continue to be delivered with sporadic funding for release time and/or instructors being provided through short term grants or student services related donations. Thus funding is not being provided in a sustainable way. This may be due to the fact that up to this point writing and speaking infused courses have not been formally endorsed as Writing and Speaking Intensive courses either by the Faculty in which they are situated or by the university administration, in spite of assiduous efforts on the part of those involved to bring this about.

Thus, in objectively reviewing the status of WAC at our institution, we must realistically identify its current situation as one of "stasis": "the state of equilibrium or ***inactivity*** [my emphasis]caused by opposing or equal forces"; or perhaps the alternate meaning might be more appropriate—" Pathology- stagnation in the flow of any of the fluids of the body, as of the blood in an inflamed area or the intestinal contents proximal to an obstruction" (http://dictionary.reference.com/browse/stasis).

Indeed Jackson and Morton (2007), revealed similar sentiments being aired at the 2006 WAC conference in terms such as "mid-life crisis", "enliven", "keeping the movement going", "staying afloat", and "in need of a tune-up". These stasis points according to Morton and Jackson may be navigated through "moves toward individual action and emergent strategies, discussing either multimodality made possible through technologies or… engaging affordances of postmodernism via constructivism, collaboration, consensus-building, and notions of both de-centering and multiplicity" (p. 48).

Others such as Hall Kells et al. (2006) have asserted that in order to attain success, developing WAC programmes need to be organic (community-based), systemic (institutionally-distributed), and sustainable (flexible and responsive). Earlier endorsement of this notion is to be found in the contentions of Miraglia and McLeod (1997) whose survey on successful WAC programmes highlighted that neither top down nor bottom up is sufficient on its own and if either one was applied without the other, failure would be the likely result. A combination of both is thus the most desirable and effective option. This is strongly supported by Condon and Rutz (2012) who assert that in order to be effective WAC programmes require a complex partnership which includes faculty, administrators, writing centers, and faculty development programs. Even Jackson and Morton (2007) in highlighting practical lessons learnt in moving beyond stasis point admit: "The effort had to centralize and decentralize, use top-down as well as bottom-up" (p. 54).

These contentions clearly point to a major limitation in our own efforts to establish WAC—the lack of full and sustainable institutional endorsement. We believe that the absence of central administrative involvement in the implementation process is related to the lack of progress in WAC implementation. This lack of support is apparent in the lack of funding and approved curricular revision to support WAC initiatives. Such involvement on the part of the administration would in no way exclude localized/bottom-up efforts such as ours as, but would in fact serve to enhance outcomes.

Furthermore, as mentioned earlier, feedback offered by those (the majority) who opted not to adopt WAC strategies or to become involved in the Communication Across the Curriculum (CAC) project, which included concerns such as additional workload, lack of support for additional grading, lack of a structure conducive to focus on writing etc. clearly demonstrate the key role of campus wide endorsement.

In addition to this, the concerns of these non participants demonstrate and concur with the view of others (Townsend, 2008; Mullin and Schorn, 2007) that while the grassroots/bottom up approach serves to initiate the transformative process, the lack of institutional/top down endorsement via the provision of resources and a formal approach to curriculum reform, often hinders the influence of this transformation and the extent to which it occurs. For instance in our situation, doubts/misgivings concerning the adoption of WAC strategies and the CAC initiative might have been allayed if the administration had put policies in place to facilitate curricular reform in the science faculty, and the rest of the campus.

Such policies would ensure the training of all staff in WAC pedagogy, and the revision of the curriculum in such a way that writing proficiency would have become a part of course objectives and integrated into learning activities, which, as has been demonstrated (Segall and Smart, 2005; Bean, 2001) is critical to the success of WAC programmes. In this way the majority of lecturers in science and other areas would be encouraged to take responsibility for their students' writing skills and to give competence in this area due weighting.

We have also come to understand that students' engagement in writing initiatives will only be realized if they understand that their writing skills are being viewed by the university and their lecturers as an integral component of their performance. In this way, they will be motivated to make a sustained effort to enhance their competence in this area. Indeed, the view that students' learning and performance is largely influenced by teaching and assessment has been attested to by (Enwistle, 2000; Wiske, 1998). This is, as mentioned before, consistent with the position of Boud (1998, p. 24) who states: "Assessment methods and requirements probably have a greater influence on how and what students learn than any other single factor".

These proposed policy decisions would not only enable the success of an institutionalized WAC programme, but they would also ensure its sustainability and vibrancy to a large extent.

Research

Research findings emanating from our Phase 1 and Phase 2 WAC/CAC initiatives were multifaceted and provided valuable insights, not only relating to the teaching of writing at our institution, but relate also generally to the areas of academic literacy and science education. In most cases, these findings were supported by the literature.

For instance, Phase 1 findings revealed that feedback on writing has a positive effect on writing performance. In the cases under study- a Preliminary Level biology course and a Level 2 (2nd year) Ecology course--- students were required to re write lab reports based on comments and re submit these. In both cases, there was and positive significant difference in performance on the redrafted submissions, and in the case of Ecology students, a significantly higher performance also in examinations on the part of those who had consistently done re drafts, compared to those who had not (McLaren and Webber, 2009; McLaren, 2014).

There was also a similar trend in a third case –a level 2 Molecular, Genetics course--- where students were encouraged to draft/redraft essays based on feedback. Results showed that those who did this consistently achieved higher examination marks, although not significantly so, than those who had produced drafts inconsistently (Boyd et al., 2008; McLaren, 2014).

Notwithstanding the fact that the results of the latter course lacked the robust statistical support yielded by the other two courses, we still considered the outcomes from this enterprise to be promising and an important contribution to the positive perception of WAC implementation.

We realise that the grading of drafts may have, in some cases, been influenced by expectations of better performance in spite of efforts to ensure the validity of the exercise, and other intervening factors may have come into play. Nonetheless we believe that the positive impact of feedback on writing performance was the strongest contributing factor to these outcomes. The contentions of other researchers concerning the influence of feedback on student writing lend support to the generalizability of our results.

Furthermore, the use of this strategy in yet another case-Analytical Chemistry—where feedback was not given in a timely or consistent way, and where there was no change in performance, adds credence to our findings.

For instance, Harris and Schaible (1997), in commenting on the use of WAC intervention strategies among biology and economics students, have concluded that improvement in writing and learning only occurs when writing is guided and written work carefully and rigorously critiqued. More recently, Vargas (2011) has confirmed that teachers' written feedback enhances coherence in essay writing and improves the general quality of student writing, and even more recently, Vardi (2012) has shown where prescriptive text-specific feedback effectively produces positive changes in student writing.

The effect of WAC/WID intervention on attitudes to writing (Phase 1) and to Speaking (Phase 2) was another major area of focus. We sought to measure changes in student attitude to writing after intervention. Results for all three courses under study revealed a significantly more negative attitude to writing after WAC/WID intervention. Conversely, a significant and positive change in Chemistry students' attitude to oral skills development practice and instruction after Speaking intervention was found and it was also revealed that the attitude of Chemistry students who had had more exposure to oral presentation skills instruction was significantly more positive compared to that of Life Sciences students who had had less exposure. There was also a strong and positive relationship between students attitude to oral skills development (Post Intervention) and perception of Instructor and method of delivery (.957; $p<.000$) (Francis and McLaren, 2014).

The findings on attitude to writing, were supported by Petric (2002) who has asserted that there is often initial resistance to writing strategies because of either stringency regarding the approach of the instructor or incompatibility of the writing approach with the cultural or institutional context. This is in turn based on Azjen's (1991) theory of planned behaviour which explains the relationship between attitudes, beliefs and behaviour. This theory posits that attitudes are determined by beliefs and thus, an attitude towards a certain behavior is determined by the belief about the outcome and the evaluation of the outcome. Even more important is the need for the individual to believe that others will be supportive of this [new] behavior and that he/she will have control over the behavior.

Given our institutional context, described in the previous section as being one where writing skills are perceived as being the sole responsibility of the English Language department by both discipline specific lecturers and students, and not an area of focus for courses outside of English Language courses, it is hardly surprising that students would not be inclined to be positive about writing.

On the other hand, speaking, which is possibly seen by students as less stringent, is more likely to elicit positive attitudes. This is partially supported by Light (2004) who asserts that students conceptualise their study and learning activities in distinct and different ways and this has important implications for both teaching and learning.

For instance, it has been found that attitudes toward oral skills training are an important antecedent to acquiring speaking skills. In support of this, Budkaew and Kessomboon (2014) found a statistically significant correlation between a positive attitude to developing oral presentation skills and performance scores in this area ($p < 0.05$) among sixth year medical students. Further to this is research on factors influencing student performance that are assumed to be relevant across domains.

Addtionally, Ho et al. (2001) report that students' perception of instruction influences teaching strategies, which in turn impacts students' approaches to learning. And according to Pintrich et al. (1993), students' views and ideas on learning play a crucial role in performance, and it is essential to take these into account to optimise learning.

Thus, we are led to conclude that there are, in the situation under discussion, untested dynamics in perception, beliefs, and attitudes as they relate to learning and attitudes, which possibly account for differences in attitudes to writing and speaking instruction.

Nevertheless, these results represent a critical addition to the existing body of knowledge on the teaching and learning of writing, and are also indicative of WAC/WID as a viable means of enhancing awareness among staff and students of the importance of writing and speaking in the disciplines, and improving writing via feedback. Even outcomes which appeared to detract from the positive impact of WAC as in the case of students' negative attitudes to writing, are being presented to the administration as clear indicators of the need for a consistent and complete integration of focused writing practices across the institution to enable student perception of congruence between lecturers' focus on writing and institutional support.

Revisioning/Reimagining the Way Forward: Possibilities and Challenges

"There is no change without dream, as there is no dream without hope" (Freire, 1994, p. 91).

We are not advancing, but returning to point zero is not an option. Our victories have been few and far between but they have not been meaningless: we have successfully transcended disciplinary boundaries and, in some way, transformed the scientific landscape and created common cause with colleagues by gaining WAC/WID adherents. They have seen the benefits of WAC strategies and are continuing to make use of these in their delivery of course content. Also, those Life Sciences and Chemistry courses selected for writing and speaking infusion, continue to function as such. We now have a part -time Writing Tutor from the English Language Section employed by Life Sciences to assist with the delivery of writing modules and to offer coaching. Thus our impact, while not being widespread, has been deep. And this is what has led us to embrace the possibilities.

For instance, the presence of a Writing Tutor in one of the science Departments, might serve as a catalyst for the other Departments who might decide to follow suit. This could in turn lead to the opportunity for establishing a satellite Writing Centre in the Science Faculty, a 'branch' of the one already existing in the Humanities. Such developments might, indirectly, foster the development and expansion of a WAC/CAC initiative across the campus. Furthermore, the presence of a new Dean of the Science Faculty, who seems sympathetic to our cause could possibly lead to stronger collaboration and funding, and most important, faculty/institutional endorsement via changes in the curriculum as they relate to the teaching and assessment of writing/communication skills. And the new Campus Principal?? Perhaps, perhaps. Yet, while continuing to embrace possibilities such as these, while 'visioning' how it could be, and seeking ways to make it so, we are being prodded by cruel realities---the reality of an administration that pays lip service to the importance of communication skills, while lacking the political will to take the requisite steps to foster competence in this area; the reality of apathetic staff members who do not wish to actively support or engage in an integrated, disciplinary and student-centred approach to developing communication skills; and the reality of dissensions among members of the institution generally, as to how scarce English language teaching resources should be deployed.

In spite of this we remain undaunted as we seek to "unveil opportunities for hope" (Freire, 1994, p. 3), as without hope "it will be hard to struggle on, and when we fight as hopeless or despairing persons, our struggle will be suicidal" (p. 3). Hope has become our strongest ally and the foundation for our dream, as we attempt to find within "the gaps and fissures" (Jackson and Morton, 2007, p. 48), of our landscape, opportunities for gaining more ground in our attempts to expand the reach of WAC/WID at our institution, beginning with the Science Faculty.

REFERENCES

Anson, C. (2006). Writing in cross curricular programs: Determining the locus of activity. *Assessing Writing.* 11(2), 100-112.

Azjen, I. (1991). Theories of cognitive self-regulation. *Organizational Behaviour and Human Decision Processes.* 50 (2), 179-211.

Bangert-Drowns, Hurley R. and Wilkinson, B. (2004). The effects of school-based writing-to-learn interventions on academic achievement: A meta-analysis. *Review of Educational Research,* 74, 29-58.

Bean, J. (2001). *Engaging ideas: The professor's guide to integrating writing, critical thinking and active learning in the classroom.* San Francisco: Jossey-Bass.

Beaufort, A. (2007). *College writing and beyond: A new framework for university writing instruction.* Utah: Utah State University Press.

Bergman, L. and Zepernick, J. (2007). Disciplinarity and transference: Students' perceptions of learning to write. *WPA Journal, 31*(1/2), 124-149.

Bocolo, P. and Mason, L. (2001). Writing to learn, writing to transfer. In: P. Tynjälä, L. Mason and K. Lonka (Eds.), *Writing as a learning tool: Integrating theory and practice* (pp. 83-104). The Netherlands: Kluwer Academic Publishers.

Boud, D. (1998, November). *Assessment and learning – unlearning bad habits of assessment.* Paper presented at Conference on Effective Assessment at University, University of Queensland.

Boyd, F., Tennant P., Persad, G., Robinson, D., and McLaren, I. (2008, Sep.). Student perceptions and outcomes of a WAC intervention exercise in an introductory biology course. *Caribbean Journal of Education,* 30 (2), 217-247.

Bransford, J., Brown, A. and Cocking, R. (2001). *How people learn: Brain, mind, experience, and school.* National Academy Press, Washington, DC.

Byrne, S. (2007). *Literacy: Why the writing is on the wall.* Retrieved February 11, 2010 from http://www.independent.ie/other/literacy-why-the-writing-is-on-the-wall-686534.html.

Caribbean Examinations Council, (2007a). *Report on candidates' work in the Caribbean Secondary Education Certificate Examination* May/June 2007. St. Michael, Barbados. 19 pp. http://www.cxc.org/SiteAssets/Jan%202007%20CSEC%20Biology.pdf (accessed Nov. 13, 2008).

Cervetti, G. N., Pearson, P. D., Barber, J., Hiebert, E. H., and Bravo, M. A. (2007). Integrating literacy and science: The research we need. In: M. Pressley, A. K. Billman, K.

H. Perry, K. E. Refitt, and J. M. Reynolds (Eds.), *Shaping literacy achievement: Research We Have, Research We Need* (pp. 157-174). New York.

Condon, W. and Rutz, C. (2012). A taxonomy of writing across the curriculum programs: Evolving to serve broader agendas. *College Composition and Communication*, 64(2), 357-82.

Duschl, R. A. and Osborne, J. (2002). Supporting and promoting argumentation discourse in science education, *Studies in Science Education*, 38(1), 39-72.

Entwistle, N. (2000, November) *Promoting deep learning through teaching and assessment: Conceptual frameworks and educational contexts*. Paper presented at TLRP Conference, Leicester.

Ellis, R. (2004). University student approaches to learning science through writing. *International Journal of Science Education,* 26(15), 1835-1853.

Emig, J. (1977). Writing as a mode of learning. *College Composition and Communication,* 28, 122-128.

Francis, T. and McLaren, I. (2014). Enhancing Communicative Competence: An Audience-Centred Approach to 'Speaking Across the Curriculum. *UWI Quality Education Forum.* (in press).

Freire, Paulo (1994). *Pedagogy of Hope: Reliving Pedagogy of the Oppressed*. New York: Continuum Publishing Company.

Ganobcsik-Williams, L., 2006. *Teaching academic writing in UK higher education: Theories, practices and models*, Houndmills: Palgrave Macmillan.

Halliday, M. A. K. and Martin, J. R. (1993). *Writing Science: Literacy and Discursive Power*. Pittsburgh: University of Pittsburgh Press.

Hall Kells, M., et al. (2006, April). *Writing across communities: A cultural ecology of language, learning, and literacy.* Paper presented at Conference of College Composition and Communication, Chicago, IL.

Hand, B., Prain V. and Yore, L. (2001). Sequential writing tasks influence on science learning. In: P. Tynjälä, P., L. Mason and K. Lonka (Eds.), *Writing as a learning tool: Integrating theory and practice* (pp. 37-56). The Netherlands: Kluwer Academic Publishers.

Herrington, A. (1985). Writing in academic settings: A study of the contexts for writing in two chemical engineering courses. *Research in the Teaching of English,* 19(4), 331-361.

Ho, A., Watkis, D. and Kelly, M. (2001). The conceptual change approach to improving teaching and learning: An evaluation of a Hong Kong staff development programme. *Higher Education,* 42, 143-169.

Jackson, R. and Morton, D. (2007). Becoming landscape architects: A postmodern approach to WAC sustainability. *The WAC Journal*, 18, 43-58.

Jerde, L. and Taper, M. L. (2004). Preparing undergraduates for professional writing: Evidence supporting the benefits of scientific writing within the biology curriculum. *Journal of College Science Teaching* 33, 34-37.

Keys, C., Hand, B., Prain, V., and Collins, S. (1999). Using the science writing heuristic as a tool for learning from laboratory investigations in secondary science. *Journal of Research in Science Teaching* 36 (10), 1065-1084.

Lerner, N. (2007). Laboratory lessons for writing and science. *Written Communication,* 24 (3), 191-22.

Light, R. (2004). *Making the most of college: Students speak their minds*. Boston: Harvard University Press.

Lillis, T. and Turner, J. (2001). Student writing in higher education: Contemporary confusion, traditional concerns. *Teaching in Higher Education,* 6 (1), 57-68.

McLaren, I., Webber, D. (2009). Writing right: Enhancing student engagement and performance in an ecology course. *International Journal of Environmental and Science Education,* 4 (4), 365-380.

McLaren, I. (2010). Another country, not my own: Crossing disciplinary boundaries, forging alliances within the framework of a Communication Across the Curriculum(CAC) initiative in the sciences. *International Journal of English and Literature* 1(2), 10-15.

McLaren, I. (2014). Contexts of engagement: Towards developing a model for implementing and assessing a writing across the curriculum programme in the sciences. *Assessing Writing*, 22, 18-32.

Miraglia, E. and McLeod, S. (1997). Whither WAC?: Interpreting the stories /histories of enduring WAC programs. *WAC Journal* 20 (3), 46-65.

Moore, R. (1993). Does writing about science improve learning about science? *Journal of College Science Teaching* 22, (2): 212-217.

Mullin, J. and Schorn, S. (2007). Enlivening WAC programs old and new. *The WAC Journal,* 18, 5–13.

National Research Council [NRC]. (1996). *National science education standards.* Washington, D.C.: National Academy Press.

National Science Education Standards [NSES] (1996). Washington, DC.: National Academy Press.

Norris, S. and Phillips, L. (2003). How literacy in its fundamental sense is central to scientific literacy. *Science Education*, 87, 224-4.

Osbourne, J. F. (2002). Science without literacy: A ship without a sail?. *Cambridge Journal of Education,* 32, 203-215.

Petric, B. (2002). Students' attitude towards writing and the development of academic writing skills. *The Writing Center Journal,* 22(2), 9-27.

Postman, N. and Weingartner, C. (1971) *Teaching as a Subversive Activity.* London: Harmondsworth, Penguin.

Pintrich, P. R., Smith, D. A. F., Garcia, T., and McKeachie, W. J. (1993). Reliability and predictive validity of the motivated strategies for learning questionnaire (MSLQ), *Educational and Psychological Measurement,* 53, 801-803.

Robertson, I. (2004). Assessing the quality of undergraduate education students' writing about learning and teaching science. *International Journal of Science Education,* 26(9), 1131-1149.

Ryden, K. C. (1993). *Mapping the invisible Landscape: Folklore, writing and the sense of place*. University of Iowa Press, Iowa City.

Segall, M. T. and Smart, R. A. (2005). In: *Direct from the disciplines: Writing Across the Curriculum* (Eds.). Portsmouth: Boynton/Cook Heinemann.

Shuman, L. J., Besterfield-Sacre, M. and McGourty, J (2005). The ABET "professional skill": Can they be taught? Can they be assessed?. *Journal of English Education.,* 94 (1), 41-55.

Thaissis, C. and Porter, T. (2010). The state of WAC/WID in 2010: Methods and results of the US Survey of the International WAC/WID Mapping Project. *College Composition and Communication.* 6 (3), 524-70.

Townsend, M. (2008). WAC program vulnerability and what to do about it: An update and brief bibliographic essay. *WAC Journal,* 19, 45-61.

Vardi, I. (2012). The impact of iterative writing and feedback on the characteristics of tertiary students' written texts. *Teaching in Higher Education,* 17(2), 167-179.

Vargas, A. J. A. (2011). *Teacher impact: Providing feedback to student essay writing regarding coherence.* Accessed July 18. From http://gistjournal.blogspot.com/2011/04/teacher-impact-providing-feedback-to.html.

Wardle, Elizabeth. (2007). Understanding "transfer" from FYC: Preliminary results of a longitudinal study. *WPA Journal, 31*(1/2), 124-149.

Wiske, M. S. (Ed.). 1998. *The Teaching for Understanding Guide.* San Francisco: Jossey-Bass.

INDEX

D

E

F

G

H

I

J

K

L

M

N

O

P

Q

R

S

T

U

V

W

Y